Tourism: economic, physical and social impacts

Tourism: economic, physical and social impacts

Alister Mathieson
Geoffrey Wall

Longman
London and New York

Longman Group Limited
Longman House
Burnt Mill, Harlow, Essex, UK

Published in the United States of America
by Longman Inc., New York

First published 1982

British Library Cataloguing in Publication Data

Mathieson, Alister
Tourism.
1. Tourist trade – Social aspects 2. Tourist trade – Economic
I. Title II. Wall, Geoffrey
338.4'791 G155

ISBN 0-582-30061-4

Library of Congress Cataloging in Publication Data

Mathieson, Alister, 1954–
Tourism: economic, physical, and social impacts.

Bibliography: p.
Includes index.
1. Tourist trade. I. Wall, Geoffrey. II. Title.
G155.A1M35 380.1'459104 81-17135
ISBN 0-582-30061-4 (pbk.) AACR2

Printed in Singapore by Singapore National Printers (Pte) Ltd

contents

list of figures

list of tables

acknowledgements

Most of the research and much of the writing for this book were undertaken at the University of Waterloo. We were very fortunate to be working at an institution with a Faculty of Environmental Studies and a Faculty of Human Kinetics and Leisure Studies in which a number of people have research and teaching interests in a wide variety of aspects of tourism and recreation. We are grateful for the willingness of many of our colleagues to share their ideas with us and to respond to our many queries. Although it may be invidious to single out particular individuals for special mention, and we would not wish to imply that they are in full agreement with what we have written, we would like to acknowledge specifically the many discussions which we have had with Ken Balmer, B. Hyma, Sally Lerner and George Priddle. Our work would have been more narrow and much less fun had we not had the benefits of their experiences.

The final version of the text was written at the University of South Carolina. The Department of Geography there provided congenial surroundings for the completion of this task and the stay there was made possible by a Leave Fellowship from the Social Sciences and Humanities Research Council of Canada. Margaret Burley typed the final manuscript with considerable efficiency and good humour.

A great deal of work is necessary to turn a manuscript into a book. This task was undertaken by many skilled people at Longman. We are grateful to them for their interest in our manuscript and for their diligence and care in seeing the project through to completion.

Alister Mathieson
Geoffrey Wall

April 1981

We are grateful to the following for permission to reproduce copyright material.

B.H. Archer for Table 3.12 from *Tourism in Gwynedd: An Economic Study* by B.H. Archer, S. Shea and R. Vane (Bangor: Institute of Economic Research, University College of Wales, 1974); E.M. Bjorklund and A.K. Philbrick for a diagram on p. 8 of 'Spatial Configurations of Mental Process' (London, Ontario: unpublished paper, Department of Geography, University of Western Ontario, 1972) as presented on p. 89 by R. Butler in *Tourism as a Factor in National and Regional Development* edited by F. Helleiner (Peterborough, Ontario: Occasional Paper No. 4, Department of Geography, Trent University, 1975); The Canadian Government Office of Tourism for an updated version of Graph B2 p. 43 from *Canadian Tourism Facts Book 1972* (Ottawa, 1974); Department of Geography, Oxford University for a figure entitled 'The Tourism-Language Change Model' in *The Social Impact of Tourism on Host Communities: A Study of Language Change in Switzerland* by P.E. White (Oxford: Research Paper 9, School of Geography, Oxford University, 1974); Department of Geography University of Waterloo for Figure 10 p. 49 from *The Environmental Impact of Outdoor Recreation* by G. Wall and C. Wright (Waterloo: Department of Geography Publication Series No. 11, University of Waterloo, 1977); G.V. Doxey for an extract from pp. 26–7 in *Heritage Canada* Vol. 2 No. 2, 1976; D.E. Lundberg for 'Distribution of the Tourist Dollar in Florida' on p. 139 of *The Tourist Business* (Boston: Cahners, 1972); The Scottish Tourist Board, and the author, R. Vaughan, for Table 11 p. 21 from *The Economic Impact of Tourism in Edinburgh and the Lothian Region* (Edinburgh: Scottish Tourist Board, 1977); The University of Wales Press, and the author, B.H. Archer, for Table 4.20 p. 40 from *Tourism in the Bahamas and Bermuda: Two Case Studies* (Bangor: Bangor Occasional Papers in Economics No. 10, University of Wales Press, 1977).

1 introduction

Tourism is the temporary movement of people to destinations outside their normal places of work and residence, the activities undertaken during their stay in those destinations, and the facilities created to cater to their needs. The study of tourism is the study of people away from their usual habitat, of the establishments which respond to the requirements of travellers, and of the impacts that they have on the economic, physical and social well-being of their hosts. It involves the motivations and experiences of the tourists, the expectations of and adjustments made by residents of reception areas, and the roles played by the numerous agencies and institutions which intercede between them.

Tourism is a luxury. Until recently, participation was restricted to the select few who could afford both the time and money to travel. Increased leisure, higher incomes and greatly enhanced mobility have combined to enable more people to partake of tourism. Improvements in transportation, the proliferation of accommodation, and the growth of inclusive tours and other forms of relatively cheap vacation travel, have further extended the opportunity to travel for pleasure. Today the majority of people in the developed world and increasing numbers in developing countries are tourists at some time in their lives. Tourism is no longer the prerogative of a few but is an accepted and accustomed, even expected, part of the lifestyles of a large and growing number of people.

Tourism is of major economic and social significance. More than 270 million tourists spend $92 billion (US) annually in places outside their own countries. This is one of the largest items in the world's foreign trade. With a world growth in visitor arrivals rate of approximately 6 per cent per annum, tourism is also one of the fastest growing economic activities. It is the most important export industry and earner of foreign exchange in many countries.

The significance of tourism has been recognized in both developed and developing countries. This can be seen in the establishment of government departments of tourism, widespread encouragement and sponsorship of tourist developments, and the proliferation of small businesses and multinational corporations contributing to and deriving benefits from the tourism industry. There is widespread optimism that tourism might be a powerful and beneficial agent of both economic and social change. Indeed, tourism has stimulated employment and investment, modified land use and economic structure, and made a positive contribution to the balance of payments in many countries throughout the world.

At the same time, the growth of tourism has prompted perceptive observers to raise many questions concerning the social and environmental desirability of encouraging further expansion. Do the expenditures of tourists benefit the residents of destination areas? Is tourism encouraging prostitution, crime and gambling? Does tourism rejuvenate the traditional arts and crafts of host cultures? Do governments direct their development priorities to satisfy the needs of tourists rather than residents? Are residents financing expensive tourist facilities through their taxes? Is tourism contributing to the destruction of the very resources which initially attracted the tourists? Are there saturation levels beyond which further growth in tourist arrivals creates more problems than benefits? What is being done to calculate these levels and to ensure that they are not exceeded?

The unprecedented growth of tourism has prompted a host of such questions, and answers are only now beginning to emerge. As tourism continues to expand, questions concerning associated economic, environmental and social effects will become more pressing. Adequate answers to such questions are predicated upon systematic and rigorous research which, until recently, has been almost totally lacking.

Research on tourism has been highly fragmented, with researchers following separate and often divergent paths. For example, there have been studies of the history of tourism (Sigaux 1966; White 1967), the development of seaside resorts (Gilbert 1939; Wall 1975), and the role of tourism in international trade (Peters 1969; Gray 1970). Impact-oriented research has been equally specialized, emphasizing specific types of impact to the exclusion of others. For instance, Doxey (1976) and Smith (1977) examined the resentments which may result from tourist–host interactions; the World Bank (1972) and Peters (1969) treated tourism as an economic phenomenon; and Hall (1970) and Cohen (1978) documented some of the effects of tourism on the physical environment. The value of such studies would be enhanced if they could be placed in a broader context.

There have been few attempts to integrate the findings of the diverse studies of the impacts of tourism, yet any assessment of the costs and benefits of tourism requires a full consideration of all the likely impacts. This volume is an attempt to synthesize the findings of research on the impacts of tourism, and to present them to the reader in a systematic fashion. The strengths and weaknesses of existing approaches will be identified, and topics which have yet to receive detailed examination in the literature will be pointed out. The materials assembled in this book are largely taken from published sources and emphasize the impacts of mass tourism in resort areas as opposed to the less apparent, scattered effects of individuals travelling in areas lacking a reputation for tourism. The value of this book will not lie in the originality of the ideas which are presented, for it will be necessary to draw heavily upon the work of others. Rather, as one of the earliest attempts to collate and access the numerous studies of impacts of tourism, it is hoped that this work will provide a balanced introduction to the topic. By drawing attention to the diversity and complexity of these impacts, conceptual and methodological difficulties will be highlighted which may be avoided in future research. An objective evaluation of the impacts of tourism is required if government agencies, planners, developers and businessmen are to appreciate the full implications of their actions. If some of the myths concerning the nature of tourist impacts can be dispelled, then the way should be open for a re-examination of the true potential of tourism as a contributor to the economic, environmental and social well-being of reception areas. Thus, it should be possible to encourage types of development which confer many of the 'blessings' of tourism without the associated 'blights'.

Organization

Widely-accepted procedures for investigating the impacts of tourism have yet to be established and few studies attempt a comprehensive examination of a broad range of types of impact. For the purposes of this work, impacts of tourism are grouped into three major categories: economic, physical and social. This distinction is somewhat artificial for, in reality, the boundaries between the categories are indistinct and their contents merge. For example, money may be spent in an attempt to reduce unacceptable environmental change. This, in turn, may have repercussions for the availability of jobs and, hence, on social well-being. Similarly, tax revenues earned as a by-product of tourism expenditures may be spent to promote more tourism, to clean up the environment or to improve social services. Ultimately, therefore, this threefold division must be justified pragmatically. However, most studies of the consequences of tourism focus primarily on only one of these three types of impact so that the organization into three major impact domains reflects the present status of research.

Just as there is overlap between the impact domains, there is also little consensus as to what should be included within them. For example, some economic studies focus upon income generation, others stress the creation of employment, whereas many reports are devoted to balance of payments questions. Such variations in emphasis hamper the comparison of findings of different investigations, frustrate the establishment of a body of theory, and contribute to inconsistency and contradiction in study conclusions.

For this study, environmental impact assessment checklists and social and economic indicator tables were consulted as a guide to the allocation of subject-matter to each impact domain. This procedure also enabled topics favoured by researchers to be distinguished from those which are largely unstudied. However, while a threefold division into economic, physical and social impacts constitutes the major organizing framework of this book, it is not rigidly imposed. There are occasions when, for the sake of clarity and in tune with the multi-faceted nature of tourist phenomena, aspects of one type of impact will be mentioned in conjunction with those of another impact domain.

Some research issues

Rapid growth of tourism has given rise to increasingly pronounced economic, environmental and social effects. However, until recently, attention has concentrated on the more obvious economic impacts with comparatively little consideration being given to the environmental and social consequences of tourism. The relative neglect of these topics has occurred in spite of increasing anxiety about environmental problems evolving from man's continued manipulation of his environment, and in spite of expanding awareness of the increasing significance of tourism. The economic emphasis of much research is a reflection of the optimism with which tourism was generally viewed in the 1960s. Great interest was expressed in the potential of tourism to contribute to economic development. Tourism was widely acclaimed as generating a multitude of beneficial effects upon such economic indicators as balance of payments, income, employment and tax revenues. While there is an element of truth in this perspective, and some types of tourism may stimulate environmental preservation and benefit residents of destination areas, contemporary tourism is on a massive scale which may pose substantial environmental and social risks.

Recognition of the size of the impacts of modern tourism has prompted a reorientation of tourism research. There has been a noticeable shift towards a more balanced perspective incorporating a critical examination of the costs, or negative impacts, of tourism. An increasing number of such studies adopt an environmental or a social perspective but the majority of such investigations are of recent vintage. The potentially serious psychological, social and cultural effects of tourism were given prominence in the seminal works of Young (1973) and Turner and Ash (1975). Following these pioneering statements, a number of authors have illuminated some of the more alarming, negative impacts of tourism (Graburn 1976; Smith 1977; Cohen 1978; Pizam 1978). These have included the modification of traditional cultures, increases in prostitution and crime, and the pollution of beaches.

The consequences of tourism have become increasingly complex and contradictory. For example, the commercialization of culture, through the marketing and sale of artefacts, may revive traditional art forms or modify them so that they are scarcely recognizable. The associated influx of money into a local economy may distort occupational stability and contribute to a breakdown in family and community cohesion. On the other hand, the commercialization of culture may lead to the creation of a 'phony folk culture' but, at the same time, create jobs and thereby alleviate existing unemployment problems. Assessments of tourism must increase in breadth and sophistication as the industry expands, and as the diversity and intricacy of impacts is magnified.

Impact studies

Evaluations of the impacts of tourism reflect the status of impact research in general. Recent environmental legislation, and demands by society for environmental impact statements for projects which significantly affect the environment, have stimulated interest in impact research, and emphasized the need for the development of sound analytical procedures. Given the varied requirements of impact assessments and the recency of their rise to prominence, it should not be surprising that there is a paucity of methodological guidelines for undertaking investigations of the impacts of tourism.

Environmental impact statements of any kind are extremely difficult to make (Wall and Wright 1977: 3–5). Five reasons for this can be highlighted. Firstly, man has been living on and modifying the earth for thousands of years so that it is extremely difficult to recontruct the environment before the intervention of man and, hence, to establish a base level against which to measure changes. In many tourist destination areas public use has existed for long periods of time so that it is now almost impossible to reconstruct the environment minus the effects induced by tourism. However, failure to establish baseline data will mean that it will be impossible to fully assess the magnitude of changes brought about by tourism.

A second difficulty concerns the problem of disentangling the role of man from the role of nature. Even without the intervention of man, the environment would not be unchanging, but would be in a perpetual state of flux. This leads to further difficulties in defining a base level. The problem is compounded because many impacts of tourism result from normal environmental processes whose actions are speeded up by the intervention of man. For instance, weathering and erosion are natural processes but they can become major problems when exacerbated by man. The processes remain unchanged but the flows of energy are radically altered.

A related problem is the difficulty of differentiating between changes attributable to pre-existing processes and changes induced by the influx of tourists. Tourism has emerged as a forceful agent of change and creates impacts which are clearly the product of tourist developments: resort landscapes, the construction of theme parks and the generation of employment and income are obvious examples. In many cases, however, it is exceedingly difficult to isolate the principal causes of change. It may be difficult to determine whether changes are directly attributable to tourist development or whether tourism is only one among a number of agents of change. In Tonga, for example, the increased demand for imported foodstuffs has resulted from an increasing population, inadequate agricultural production to feed the people, and demands imposed by international tourism. The extent to which tourism has contributed to the deep-seated social and economic problems emanating from this situation is not accurately known. Tourism, undoubtedly, has been one important contributing factor but it may also be a highly visible scapegoat for problems which already existed prior to the advent of modern tourism. It certainly is easier to blame tourism than it is to address the conditions of society and environment.

Thirdly, the complex interactions of tourism phenomena make total impact almost impossible to measure. Many of the impacts of tourism are manifested in subtle and often unexpected ways. In other words, primary impacts give rise to secondary and tertiary impacts and generate a myriad of successive repercussions which it is usually impracticable to trace and monitor. Cross impacts are a direct result of the interactive nature of economic, environmental and social phenomena. Interactions between components of each of these impact domains induce further changes which reverberate through the system, creating a complex array of impact flows. For example, organized safaris to national parks in Africa can modify the feeding and breeding habits of wildlife. Preservation of wildlife may be in the interests of tourism developers and may have economic benefits to countries endowed with this resource. However, the establishment of national parks has forced the rapidly growing African population to farm in areas of low fertility where agricultural production may be barely sufficient to feed the population. Special impacts occur to particular groups of people, such as racial or cultural minorities, or to unique types of wildlife or vegetation. The distinctive characteristics of such groups can make them extremely attractive to the tourist but also highly vulnerable to impacts from tourist activity.

A fourth obstacle stems from spatial and temporal discontinuities between cause and effect. For instance, erosion in one location may result in deposition elsewhere, destruction of key elements of an animal's habitat may lead to population declines throughout its range, or development of new facilities may divert tourist traffic away from existing locations. A considerable time may elapse before the full implications of an activity are apparent. Thus, there are great difficulties in establishing both temporal and spatial constraints for undertaking impact studies.

A fifth methodological issue is the selection of impact indicators. What indicators should be used and what do they really mean? For example, what is the significance of prostitution and an increased crime rate when compared with traffic congestion or an expanded tax base? In other words, there is a problem in the identification of which variables best indicate the changing situation and, in consequence, of what to measure. A related problem is the assigning of weights to the selected indicators, as indicators of impact vary in their significance to the impacted system, and devising means of combining disparate measures into a composite index of the magnitude of impact.

Problems, such as those which have been discussed above, have restricted the scope and accuracy of research results and have encouraged investigators to narrow the focus of their research. There has been a tendency to examine impacts from selected, specific developments or projects in isolation from the broader tourist phenomena of which they are a part; to concentrate on primary impacts to the exclusion of secondary and tertiary impacts; to measure the more tangible, quantifiable impacts, such as economic impacts, to the neglect of the less readily measured social and environmental impacts; and to stress positive impacts or benefits, and to overlook undesirable consequences or costs. Future examinations of the consequences of tourism for host communities will continue to be concerned with the types and magnitude of impact and whether they are predominantly beneficial or costly to the destination area. In addition, it will be necessary to give greater attention to the scale of impact for the spheres of influence of tourist development vary and their consequences may be viewed differently depending upon whether they are assessed from a local, regional, national or international perspective. At the same time, it will be appropriate to give greater attention to the assessment of who gains and who loses. Costs and benefits of tourism are not evenly distributed. What may be a benefit to one group or individual within a community may be a cost to the neighbours. Investors in tourist developments and associated service industries may gain at the expense of other residents of the destination area who may suffer increased crowding, congestion, noise, pollution and modified lifestyles. Furthermore, tourism is dynamic and impacts and their significance are constantly changing due to modifications of the goals of both the tourists and their hosts, fluctuations in the processes shaping the economic and physical environments, and technological changes and other developments in the tourist industry itself. This implies that impacts will change through time and periodic monitoring may be desirable.

Relationships between leisure, recreation and tourism

Discussions of recreation and tourism are plagued by imprecise terminology. Although there is a considerable body of literature which attempts to clarify the meanings of such terms as recreation and leisure, universally acceptable definitions have yet to be derived. The problem is compounded by the indiscriminate use of words such as pleasure, fun, spare time and enjoyment which are often used as synonyms for recreation and leisure and as substitutes for each other. In the interests of clear thinking it is desirable to make a distinction between the meanings of leisure and recreation. Leisure can be regarded as a measure of time: it is the time remaining after work, sleep and necessary personal and household chores have been completed. It is the time available for doing as one chooses. Leisure may thus be defined as 'discretionary time'. Recreation embraces the wide variety of activities which are undertaken during leisure. Outside of professional circles, there has probably never been one word or phrase in common circulation to describe that time which we think of as leisure. People talk about concrete, discrete activities, such as watching television, skiing or going to the cottage, and not about tourism, recreation or leisure. Leisure, recreation and tourism are abstractions from common experience, abstractions which only those who stand aside from that experience can perceive. The language is that of the academic and the planner rather than the participant (Cunningham 1980).

The simple distinction between leisure as discretionary time and recreation as activity is difficult to implement, for many activities include both obligatory and discretionary components. For instance, without food we would die and eating is a necessity; it is also a popular form of recreation from which many people derive great pleasure. Similarly, gardening and attending conventions are activities which can be both enjoyable and a chore. Such difficulties have prompted some authors to argue that leisure and recreation are states of mind and that they are best defined in psychological terms (Driver and Tocher 1974). While one can be sympathetic to this viewpoint and can acknowledge that individuals recreate for a wide variety of reasons and may even derive different satisfactions from the same activity, psychological definitions have their own inherent difficulties. The designation of areas for excitement, danger or relaxation is uncommon among recreation and tourism planners and site managers who usually operate on the basis of activities, designating areas for camping, skiing or hunting. However, the psychological definitions do serve to remind us that opportunities to recreate are not provided in and for themselves; they are made available to enable participants to achieve a wide range of satisfactions.

Tourism, recreation and leisure are not the prerogative of any one discipline. Recreations in the home, such as reading and watching television, are probably best studied by sociologists and psychologists, although economists may be interested in associated spending patterns. While recognizing that other disciplines have important roles to play, it is suggested that the geographer is in a position to make a distinct and significant contribution to the understanding of tourism and outdoor recreation. Tourism and outdoor recreation are land uses. They are in competition with agriculture, forestry, mining, housing, industry and a variety of other functions for the same scarce resources of land and water. Tourist and recreation facilities such as ski areas, resorts, parks and swimming pools have service areas comparable to those of stores or ports, and tourism and recreation create patterns of movement analogous to those associated with commuting or migration and susceptible to analysis by similar methods. From these examples alone it should be evident that the concepts and methods of the geographer are appropriate to analyses of recreation and have the potential to further the understanding of recreational phenomena.

Tourism and outdoor recreation have two basic aspects: the supply of facilities and the demand for participation. Supply and demand interact to produce the pattern of tourism and outdoor recreation, which may be defined as the spatial and temporal incidence of tourism and outdoor recreation. These patterns have associated economic, environmental and social impacts and give rise to planning and management problems and opportunities.

Interactions between supply and demand occur at a variety of scales reflecting the time available for outdoor recreation and the distances that can be traversed during that time. When only short periods of time are available, as, for example, in the evening, recreation, of necessity, takes place in or relatively close to the home. In contrast, during vacations, when several days of leisure may be juxtaposed, it is possible to travel long distances. There is thus a continuum from recreation in the home to recreation at considerable distances from the home base, the latter often being associated with the acquisition of temporary accommodation. It follows that tourism can be regarded as an extreme form of recreation which is distinguished by relatively long lengths of stay away from home and relatively large distances travelled (Britton 1979). The literature generally focusses on either recreation or tourism but, it is argued, they are aspects of the same phenomenon which can usefully be considered together: after all, recreationists and tourists may be found together at the same sites doing similar things.

The existence of the recreational time–distance continuum draws attention to the fact that the temporal distribution of leisure may be as significant as its quantity. If, for example, the working week were reduced by several hours, it would make a great deal of difference to patterns of recreation if these hours were distributed evenly across the week, added to the weekend, or accumulated towards a longer vacation. Other things being equal, the larger the size of the unit of leisure, the smaller are the distance constraints, and the greater is the freedom of locational choice of the potential participant. However, such time–distance relationships are further modified, particularly on long journeys, by the availability of money, for wealthy travellers may increase their time at a destination by substituting fast but expensive air travel for cheaper but slower ground transportation.

The difficulties of distinguishing between tourism and other forms of recreation have been recognized by most recreation and tourism texts. It is difficult to isolate the activities and demands of tourists as opposed to those of participants in other forms of recreation. Tourism and recreation often share the same facilities and compete for space and finance: facilities, such as theme parks, may be established to attract tourists and also to cater for recreationists; local demand for new recreational facilities (for example, artificial ski slopes) may be prompted by experiences gained as tourists abroad; measures adopted to improve the environment and to conserve and restore national park landscapes and historic monuments benefit both recreation and tourism. The demands and effects of recreation and tourism are, therefore, closely interrelated. Burkart and Medlik (1974: 10) described the confusing situation as follows:

> tourism represents a particular use of leisure time and a particular form of recreation but does not include all uses of leisure time nor all forms of recreation. It includes much travel but not all travel. Conceptually tourism is, therefore, distinguished in particular from related concepts of leisure and recreation on the one hand, and from travel and migration on the other.

Attempts to differentiate between recreation and tourism on the basis of motivations, activity types, modes of travel and distances travelled have met with mixed success. One way of classifying tourism and recreation is through the examination of the availability and use of leisure. According to Lawson and Baud-Bovy (1977: 4) leisure is of four types:

1. Daily recreation uses facilities in close proximity to one's home or place of work and for short periods during the day;
2. One-day recreation encompasses excursions to the fringes of urban areas or further into the countryside but within easy reach of home. No overnight stay is required;
3. Weekends and short holidays may be spent with some frequency relatively close to one's residence in second homes or other temporary accommodation;
4. Long holidays involve fewer distance constraints and may be taken in one's own country or abroad.

From this perspective, tourists would be included in the 'long holiday' and 'weekends and short holiday' classifications, although not all people assigned to these groups would, of necessity, be tourists.

Tourism, then, is but one of a range of choices or styles of recreation expressed either through travel or a temporary short term change of residence. Tourism, on its modern scale, is a relatively new use of leisure. Marked and rapid changes in technology and in social, political and economic systems have enabled people to pursue new and different forms of recreation and have magnified the importance of tourism. Tourism is an evolutionary development in the use of leisure and represents an expanded opportunity for the exercise of choice in the selection of recreational activities.

Definitions of tourism

Before one can examine tourist phenomena and assess their effects on economic, physical and social environments it is necessary to devise appropriate definitions. Frechtling (1976: 59) stated that definitions for travel research should:

1. Be discrete and unambiguous and must clearly define one activity or entity as distinct from all others, i.e. there should be no confusion over what is included in or excluded from a category;
2. Facilitate measurement as much as is consistent with other objectives;
3. Follow established usage as closely as possible. In other words, in developing definitions reference should be made both to major travel studies and to everyday language. This should facilitate comparison of results with those of other studies, aid continuity in research and permit a cumulative body of knowledge to be developed.

The above principles have been poorly adhered to in the research which has been undertaken to date. In fact, there are almost as many definitions of tourism as there are studies of the phenomenon (Cohen 1974). A survey of eighty travel and tourism studies conducted by Frechtling (1976) yielded forty-three different definitions for the three terms of traveller, tourist and visitor. Such results indicate the lack of coordination in travel research and hamper comparisons between travel research data.

In assessing the impacts of tourism, it is fundamental to define the major components of tourism: the tourist. 'Tourist' is derived from the term 'tour' which, according to *Webster's International Dictionary* (1961: 2417), means: 'a journey at which one returns to the starting point; a circular trip usually for business, pleasure or education during which various places are visited and for which an itinerary is usually planned'. Accordingly, the *Oxford English Dictionary* (1933: 190) defines the tourist as: 'one who makes a tour or tours; especially one who does this for recreation; one who travels for pleasure or culture, one who visits a number of places for their objects of interest, scenery or the like.'

Dictionary meanings of the term 'tourist' have been expanded and complicated with the rise of tourism research. Ogilvie (1933) was one of the first to incorporate additional meaning through use in social science research. He described a tourist as any person whose movements fulfil two conditions:

1. That the person's absence from home was for a relatively short period;
2. That money spent during absence is money derived from home and not earned in the destination visited.

Cohen (1974: 529) commented that Ogilvie's definition: 'translates the contention found in other definitions, namely that the tourist is a traveller for recreation or pleasure, into economic terms: the tourist is, economically speaking, a consumer and not a producer.'

In his analysis of definitions, Frechtling (1976: 60) outlined four basic criteria used in their formulation:

1. Purpose of trip;
2. Mode of transportation used;
3. Length of stay;
4. Distance travelled.

It is generally agreed that the former two criteria are insufficient for practical contemporary definitions and attention has been concentrated on the latter two criteria. Length of stay is a principal component of the United Nations definition which requires that tourists stay in excess of twenty-four hours but less than twelve months. The United Nations definition will be discussed in more detail below. Some definitions are dominated by a distance criterion. For example, the National Tourism Resources Review Commission (NTRRC) defined a tourist as: 'one who travels away from his home for a distance of at least 50 miles (one way) for business, pleasure, personal affairs or any other purpose except to commute to work'. The US Census Bureau has settled for a similar definition but has extended the minimum distance to 100 miles.

Cohen (1974) has also reviewed the literature which attempted to define the 'tourist'. He identified six major dimensions: permanency, voluntariness, direction, distance, recurrency and purpose. He defined the tourist as: 'a voluntary, temporary, traveller, travelling in the expectation of pleasure from the novelty and change experienced on a relatively long and non-recurrent round-trip' (Cohen 1974: 533). This definition has the merits of being both concise and comprehensive but for the collection of data it is necessary to be even more explicit and precise time and distance constraints must be established.

In 1963 the United Nations sponsored a conference on travel and tourism in Rome. The conference recommended definitions of 'visitor' and 'tourist' for use in compiling international statistics. For statistical purposes the term 'visitor' describes any person visiting a country other than that in which he has his usual place of residence, for any reason other than following an occupation remunerated from within the country visited. This definition covers:

1. Tourists who are temporary visitors staying at least 24 hours in the country visited and the purpose of whose journey can be classified under one of the following headings;
   (a) leisure (recreation, holiday, health, study, religion and sport);
   (b) business, family, mission, meeting;
2. Excursionists who are temporary visitors staying less than 24 hours in the country visited, including travellers on cruise ships (International Union of Official Travel Organizations [IUOTO] 1963: 14).

In 1968 IUOTO (now the World Tourism Organization) approved the 1963 definition and has encouraged countries to use it. Leiper (1979: 393) has noted that one consequence of this definition is that statistical data on international tourists include trips for purposes beyond the popular use of the word. For example, most people do not consider business trips as tourism. Nevertheless, the United Nations terminology has received widespread acceptance and, in the context of this book, it has proved useful in locating literature specifically pertaining to tourism. According to the United Nations definition, a tourist may be classified, in the terminology of Lawson and Baud-Bovy, in both 'weekend and short holiday' and 'long holiday' recreation. 'Daily' and 'one day recreation' participants can be grouped under the 'excursionist' category. However, this differentiation fails to distinguish between impacts of tourism as opposed to impacts of other forms of recreation because both leisure groups may be participating in similar activities at the same locations. As contemporary tourism is a mass phenomenon and is highly concentrated in particular destinations, its effects are likely to be more pronounced than those of excursionists, although the impacts of the latter are likely to be very similar in kind.

Other terms which require definition are 'international tourist' which include those individuals travelling across an international border and who remain away from home for at least twenty-four hours, and domestic tourists who are those individuals travelling within their own country but who remain away from home in excess of twenty-four hours. Statistical definitions of the tourist in a domestic setting (travelling within the country of residence) have varied among countries and regions, but have generally included three major elements: distance, purpose of travel, and length of stay.

Since the focus of this book is on the impacts of tourism which accrue to destination areas, it is necessary to define the domain in which impacts occur. A 'destination area' is a place having characteristics which are known to a sufficient number of potential visitors to justify its consideration as an entity, attracting travel to itself, independent of the attractions of other locations. The natural and man-made features, infrastructural characteristics, economic structures, and the attributes of the host populations of destination areas are of interest here.

Commentary

The objective of this book is to assess the impacts of tourists on destination areas. The book is partly descriptive in that it presents some of the viewpoints and findings of other authors, partly remonstrative in that it criticizes the approaches and methods frequently employed in tourist impact research, and partially cautionary as it draws attention to the lessons to be learned from existing tourist development and discusses possible means for their amelioration. The book is divided into six major chapters. Following the introduction, the second chapter outlines conceptual frameworks for the tourist phenomenon and tourist decision-making. It is suggested that the magnitude and types of tourist impacts are outcomes of these frameworks. The next three chapters constitute the major part of the book and successively examine the economic, physical and social impacts of tourism. Each of these chapters examines relationships between tourism and components of the sub-system in question. The nature of effects and their implications are discussed and policy implications are examined. The final chapter presents broad conclusions derived from the preceding discussions and presents observations concerning the objectives of tourist planning.

2 conceptualization

Tourism is a multi-faceted phenomenon which involves movement to and stay in destinations outside the normal place of residence. A conceptual framework of tourism is presented below. This framework emphasizes some of the major components of tourism and also places the impacts of tourism into a broader context.

Tourism is composed of three basic elements:

1. A dynamic element which involves travel to a selected destination or destinations;
2. A static element which involves the stay in the destination;
3. A consequential element, resulting from the two preceding elements, which is concerned with effects on the economic, physical and social sub-systems with which the tourist is directly or indirectly in contact.

Thus, tourism is a composite phenomenon which incorporates the diversity of variables and relationships to be found in the tourist travel process. Some of the major variables and their interrelationships are presented in a conceptual framework, below (Fig. 1).

Impacts of tourism are viewed as being more than the results of a specific tourist event or facility. Impacts emerge in the form of altered human behaviour which stems from the interactions between the agents of change and the sub-systems on which they impinge.

Certain reservations are in order in considering the conceptual framework. It has not been devised as a tool for predicting demand nor for developing strategies to promote and market tourist products. The purpose of devising a framework, as has already been stated, is to illuminate tourism as an amalgamation of phenomena and their interrelationships. All variables within the framework will not assert the same degree of influence (i.e. have equal weighting) nor can they all be readily quantified. Nevertheless, the framework has sound conceptual assumptions from an impact perspective because it explicitly recognizes:

1. Sets of variables and their interrelationships and the ways in which they influence the nature, direction and magnitude of tourist impacts;
2. That impacts linger and interact with each other;
3. That impacts operate continuously but they change through time with changing demands of the tourist population and with structural changes in the tourist industry itself;
4. That impacts result from a complex process of interchange between tourists, host communities and destination environments;
5. That assessment of impacts should encompass all phases of the travel ex-

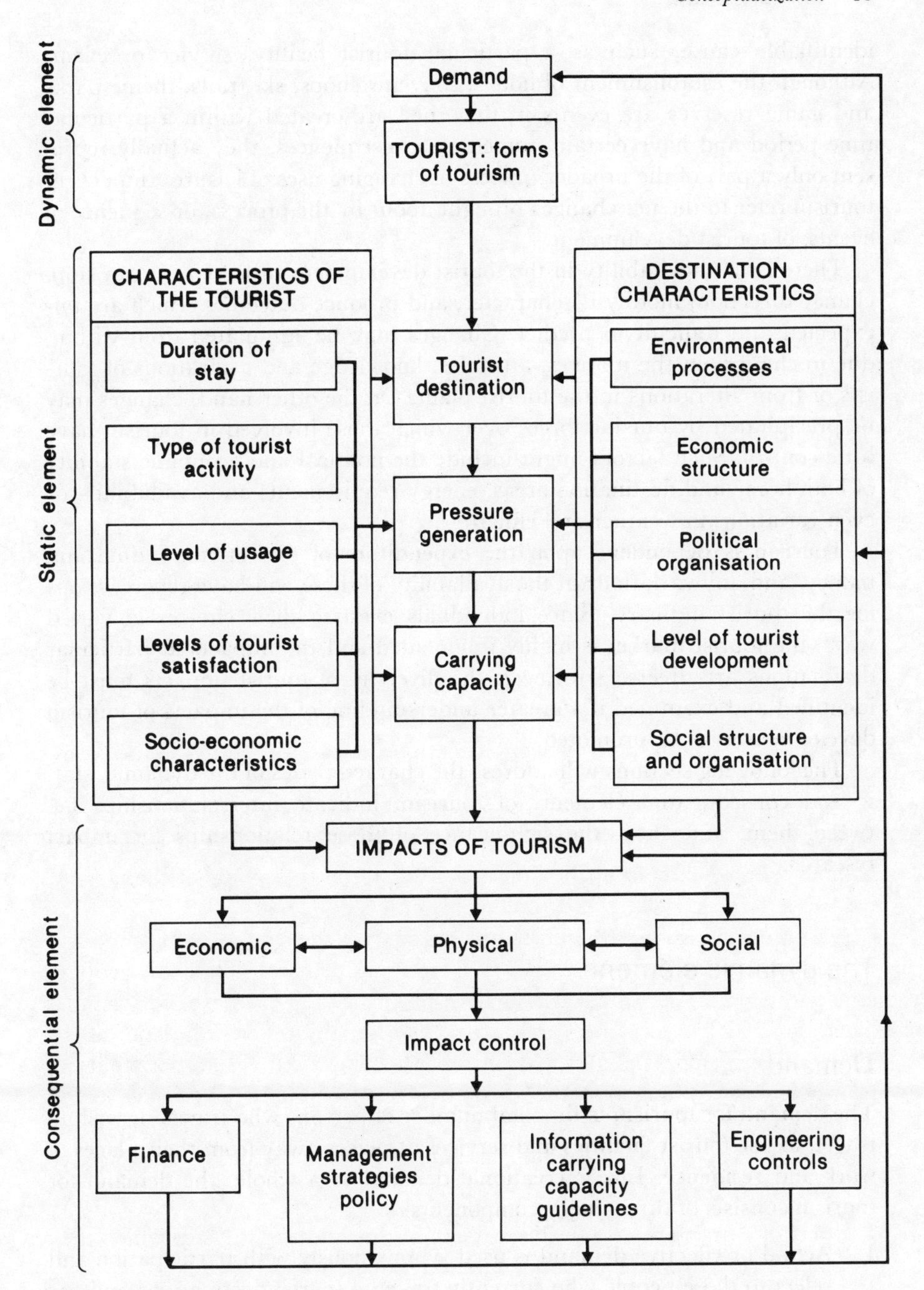

Fig. 1 A conceptual framework of tourism

perience, including initial preparations, the journeys to and from the destination, the stay, and post-trip recollections.

The framework also recognizes that impacts result from processes of change. Impacts of tourism are not point events which result from a specific, identifiable cause, such as a particular tourist facility, service or event. Although the establishment of hotels, souvenir shops, ski trails, theme parks and game reserves are events in that they are created within a particular time period and have certain immediate consequences, they actually represent only a part of the broader process of changing uses of leisure. Impacts of tourism refer to the net changes brought about by the process, or sequence of events, of tourist development.

There is no inevitability in the tourist development process as it can stop, change direction, modify its character, and produce outcomes which are unexpected and difficult to predict. Changes may be stimulated from within, due to changes in the motives, attitudes, knowledge and aspirations of tourists or from alterations in the tourist plant. On the other hand, changes may be precipitated by outside forces over which those involved in tourism have little control. Such factors might include the political and economic stability of both home and destination areas, energy requirements and availability, or even variations in weather and climate.

Tourism is dependent upon the expenditure of discretionary time and money, and any variations in the availability of these will have repercussions for the tourist industry. Since individuals exercise their choices in varied ways, the tourist market is highly fragmented and the impacts at particular destinations are diverse. However, the diversity of tourist impacts must be identified and examined if a greater understanding of the impacts of tourism development is to be promoted.

The following sections will address the characteristics of the dynamic, static and consequential elements of tourism, indicate interrelationships between them, and show the significance of those relationships for impact research.

The dynamic element

Demand

The demand for tourism is the total number of persons who travel, or wish to travel, to use tourist facilities and services at places away from their places of work and residence. Like recreational demand as a whole, the demand for tourism consists of three major components:

1. Actual or effective demand is used synonymously with participation and refers to those people who currently travel to tourist destinations and use their services and facilities;
2. Potential demand includes those persons motivated to travel but who are unable to do so because of temporal or financial constraints;
3. Deferred demand includes those people who could travel, if motivated, but they do not do so because they either lack the knowledge of opportunities, facilities, or both.

Potential and deferred demand may be considered together as suppressed demand.

The subject of demand has received considerable attention in the literature on tourism but, unfortunately, the terminology is often used imprecisely. Suppressed demand is extremely difficult to measure and, in consequence, is often ignored. Authors frequently use the word demand when they are considering only effective demand, participation or consumption. Great care, therefore, should be taken in interpreting the literature on the demand to participate in tourism.

Many writers have noted economic influences on demand (Peters 1969; McIntosh 1977; Schmoll 1977), and others have attempted to devise ways in which demand can be predicted (Quandt 1970). Tourism, being a particular form of recreational behaviour, is influenced by the same determinants of demand as outdoor recreation. These have been well documented in the literature and are predominantly economic, social and technological in nature. They include:

1. Rising per capita incomes, higher purchasing power and greater disposable incomes which have enabled people to engage in more expensive forms of recreation, including tourism, and have enabled them to travel greater distances from their homes;
2. The desire to escape the pressures of everyday urban living, to experience a change of environment, and to seek leisure outlets beyond the limits of the city;
3. Increased mobility. Sophisticated developments in transportation, particularly the automobile and the aeroplane, have made travel easier, faster and more comfortable. This has meant that distant tourist destinations have become more accessible to larger numbers of people;
4. Higher levels of education. This has increased the desire of people to see and experience new things, people and places, and has also increased their awareness of the means of doing so. A liberalization of attitudes towards leisure has been associated with this trend so that the ability to travel is now considered to be an integral part of modern lifestyles and consumption patterns.

Changes in the above general determinants of demand have enabled more people to spend their leisure more freely, and have been responsible for the rapid growth of tourist arrivals by extending the opportunity to travel.

Increased tourist demand has also been stimulated by a number of more specific factors. These include:

1. The development and increasing use of travel intermediaries, particularly travel agents and tour operators. This has taken away much of the planning responsibility from the tourist;
2. The growth of chartered air travel;
3. The development of new tools to sell the tourist product. The extensive use of the package tour has been particularly noteworthy. The marketing of travel, accommodation, food and sightseeing opportunities as a single

product has eased the planning concerns of the less experienced traveller, reduced costs through the acquisition of group rates and, thereby, greatly increased the volume of effective demand. At the same time the growth of the package tour has tended to concentrate tourists into specific localities. These include Hawaii and the coasts of Spain, Florida and the Caribbean;

4. Greater organization within the tourist industry. This has been achieved through the efforts of travel intermediaries, international tourist organizations, national and regional tourist boards, and the vertical and horizontal integration of companies involved in the travel and accommodation sectors.

The above factors, collectively, have stimulated the growth of the tourist industry. They have enabled suppressed demand to be converted into effective demand. People are able to travel further and faster, more frequently and more cheaply. This has increased the volume of tourist traffic, stimulated the creation of new destinations, and concentrated people into existing resorts where saturation levels may have been reached or exceeded.

Forms of tourist travel

Tourism is a generic term which encompasses many different types of travel. The forms of tourism contribute to the nature of the experiences gained by tourists and influence the effects which they have on the host society and their environment. Typologies of tourism have been proposed by Cohen (1972: 167–8), Wahab (1975: 10) and Smith (1977: 2–3). Wahab listed several different forms of tourism according to a wide range of variables. These included:

1. Sex – masculine and feminine types of tourism;
2. Means of transport – tourism by land, sea or air;
3. Geographic locality – international and domestic tourism;
4. Price and social class – de luxe and middle-class tourism;
5. Age – youth and adult tourism.

Typologies based on purpose of trip have been established by Wahab (1975: 10) and by Smith (1977: 2–3). Wahab produced a fivefold classification which included recreational tourism, cultural tourism, health tourism, sport tourism and conference tourism, although distinctions between the categories are not always clear. Smith's typology included Wahab's first two categories plus historical tourism, ethnic tourism and environmental tourism. Cultural tourism, for example, informs tourists about other countries and peoples, their lifestyles, customs and languages. Historical tourism encompasses visits to ancient monuments, archaeological digs, and other places of historical interest such as museums and battlefields. Again, there is overlap between components of the typology and, in many cases, the semantic differences do not appear to indicate real differences in structure or content.

The typologies of both Wahab and Smith assume that tourists travel to destinations for a narrow range of specific reasons. However, motivational research has shown that tourists may choose a destination for more than one reason and their behaviour may not entirely reflect their initial travel motivations. The categories of these typologies do not adhere closely to what tourists perceive as their travel motivations, or with their activities as indicated in the literature. For example, Smith's division of cultural tourism includes considerably more than the understanding of the vanishing lifestyles of traditional societies. As Ritchie and Zins (1978: 257) noted, cultural tourism also includes the acquisition of knowledge about peoples of contemporary societies, observed through art and crafts, work, religion, language, traditions, food and dress. Categories based on single travel motivations do little more than indicate very general tendencies in the choice of travel destinations. They do not explain the nature of tourist phenomena nor their manifestations.

Cohen (1972: 167–8) suggested a classification of tourists based on the assumption that tourist experiences combine varying degrees of novelty with an element of the familiar, the excitement of change mixed with the security of accustomed habits. Cohen's typology recognizes the importance of individual motivations and also assumes that the extent to which familiarity and novelty is experienced is influenced by tourist preferences and the institutional setting of the trip. He developed a fourfold classification of tourist experiences and roles as follows:

1. *The organized mass tourist.* This role is typified by the package tour in which itineraries are fixed, stops are planned and guided, and all major decisions are left to the organizer. Familiarity is at a maximum and novelty at a minimum;
2. *The individual mass tourist.* In this role, the tour is not entirely planned by others, and the tourist has some control over his itinerary and time allocations. However, all of the major arrangements are made through a travel intermediary. Like the organized mass tourist, the individual mass tourist remains largely within the 'environmental bubble' of home country ways and mixes little with members of the host community. Familiarity is still dominant.
3. *The explorer.* Explorers usually plan their own trips and try to avoid developed tourist attractions as much as possible. In spite of the desire to mix with members of the host community, the protection of the 'environmental bubble' is still sought. Novelty now dominates but the tourist does not become fully integrated with the host society.
4. *The drifter.* Drifters plan their trips alone, avoid tourist attractions and live with members of the host society. They are almost entirely immersed in the host culture, sharing its shelter, food and habits. Novelty is dominant and familiarity disappears.

Each of the four tourist roles represents one stereotypical form of tourist behaviour. Because each role is associated with tourist groups of different sizes and with different degrees of integration with residents of the host community, the classification has implications for the impacts of tourism. The impacts of tourism considered in this book are primarily the results of institutionalized tourism, or the combination of Cohen's first two categories. Little is known of the effects of interactions of host society members with non-institutionalized tourists, i.e. explorers and drifters.

Tourists of the institutionalized type have been described as 'sightseers' in a later article by Cohen (1974: 544). Sightseers normally only visit a destination once and visit a variety of places within the confines of one trip. The emphasis of sightseers is strong on travel and weak on sojourn. This tends to be the opposite of non-institutionalized tourists whom Cohen (1974: 544) called vacationers.

Cohen's typology has more apparent implications for destinations than that of Wahab or Smith. Institutionalized, mass tourism imposes considerable demands for the supply of facilities and services with which the tourist can readily identify. The standardization of facilities, the transformation of natural attractions, and the construction of artificial ones produces an 'ecological bubble' of the tourist's accustomed environment. The claim of the Holiday Inn chain that patrons of their establishments will have 'no surprises' caters to the conservatism of many mass tourists. The development of similarity, even uniformity, in the tourist experience has important economic, cultural and infrastructural effects on destinations. It tends to encourage a homogeneity in tourist landscapes. The extent to which the roles of tourists and hosts are predefined and social expectations are made known also largely determines the manner in which tourists interact with members of the host society. In mass tourism, social contacts tend to be both limited and superficial.

Similar types of tourism will have different types of impact depending upon the nature of the societies on which they impinge. Although Cohen's typology emphasizes the motivations and travel arrangements of tourists, it also implies that destinations respond differently in accordance with different forms of tourism. Such typologies indicate that it is inadequate to discuss tourists and destinations in general terms. Because they encourage a consideration of particular types of tourist and particular attributes of destinations, such typologies may assist in the quest for informed decisions regarding the allocation of resources within the tourist industry and the desirability of extending facilities and services within destination areas.

This brief discussion of the dynamic element of the conceptual framework has identified some of the factors which have contributed to the growth of tourism. It has also examined different types of tourist travel arrangements and indicated some of their implications for the impacts of tourism. However, impacts of tourism are not restricted in their causes to the dynamic elements which have been discussed. They are the outcome of both the dynamic and static elements. They are the result of the interaction of the tourists with the destination area and its residents.

The static element

Carrying capacity

Carrying capacity is the maximum number of people who can use a site without an unacceptable alteration in the physical environment and without an unacceptable decline in the quality of the experience gained by visitors. There is now an extensive literature on the carrying capacity of recreational areas but this is not the place to review it for this has been done adequately elsewhere (Mitchell 1979: 176–200). However, several points merit emphasis. The concept appears to be applicable to both natural and man-made environments, although it has been largely applied to the former. There are few examples of the application of the concept in studies of tourism although there are signs that this may be changing (McCool 1978). A recreation site or tourist resort, be it natural, man-modified or man-made, does not have one set carrying capacity. The capacity will reflect the goals established for the site or resort, and these should specify the level of environmental modification which is unacceptable and the nature of the experiences to be provided. It may be influenced by such factors as capital availability and managerial experience. Carrying capacity remains an elusive concept, but the time when researchers and managers sought one mythical magic number, which could be approached with safety but exceeded at peril, have passed. Discussions of the carrying capacity concept have directed attention to the need for clear and precise statements of goals, and assessments of the extent to which those goals are being realized.

Measures of capacity in tourist areas can relate to approaching routes, to the resort area and its attractions as a whole, or to individual services and facilities (Hall 1974: 392). In each of these cases capacity has economic, physical and social attributes which may be measured. Each capacity type will vary for different destinations depending upon the physical characteristics of the destinations, the types and levels of use, and the goals which they are expected to satisfy. Each type of capacity will have a tolerance limit for each destination or facility. This limit marks a threshold of change beyond which tourist activity induces effects which are primarily negative. If tolerance limits are not exceeded then the effects of tourism will be generally positive.

Capacities exist for each of the economic, physical (or environmental) and social sub-systems of the destination. Finding space for souvenir shops, restaurants, hotels and car parks may mean the appropriation of land occupied by shops and specialist functions catering primarily to the local market. Thus the notion of economic carrying capacity – the ability to absorb tourist functions without squeezing out desirable local activities – can be developed. The wear and tear of historical buildings and the contamination of beaches by untreated sewage are examples of physical carrying capacity being exceeded. It is well known that host peoples' levels of tolerance for the presence and behaviour of tourists has been surpassed in some locations and, thus, the social carrying capacity has been overreached.

Although there have been few attempts to measure the carrying capacities of resort areas, the concept is of great importance in the determination and understanding of tourist impacts. It also has implications for the planning of destination areas and for the regulation of visitor activities. As Wagar (1964: 21) has pointed out, carrying capacity is not an end in itself, but is a means to an end. Changes to tourist destinations are inevitable but application of the concept of carrying capacity has the potential to indicate the degree and direction of change and to aid in the assessment of the extent to which such changes are acceptable.

Capacity levels are influenced by two major groups of factors:

1. The characteristics of the tourists;
2. The characteristics of the destination area and its population.

Tourist characteristics

The characteristics of tourists having implications for carrying capacity and the nature of tourist impacts include the following:

1. The socio-economic characteristics of visitors. These include age and sex profiles; income levels; availability of spending money; motivations, attitudes and expectations; perceptions of resource quality; racial and ethnic backgrounds and overt behaviours. Together, these variables contribute to interpersonal style;
2. The level of use. Number of visitors, their distribution in space and time and, hence, visitor density are of obvious importance;
3. The lengths of stay;
4. The types of tourist activity;
5. The levels of tourist satisfaction.

Each of the above visitor characteristics is important because they all influence the magnitude, frequency and kind of interaction with the physical attributes of the destination and its people.

Characteristics of the destination area and its population

Characteristics of the destination area and its residents that influence carrying capacity and tourist impacts include:

1. *Natural environmental features and processes*. These include topography; mountains, lakes, rivers and sea; soil, vegetation, flora and fauna; sunshine, temperature, precipitation, photosynthesis, erosion, and other environmental processes;
2. *Economic structure and economic development*. This includes the level of economic development; the diversity of the economic base; the spatial characteristics of development; patterns of investment; and the import-export characteristics of the revenue of the destination;

3. *Social structure and organization.* This category includes the demographic profile of the host population; the strength of local culture; availability and quality of social amenities; patterns of social organization; women in the work-force; religious affiliations; moral conduct; levels of health and safety; perceptions, attitudes and values towards tourists; language; traditions and gastronomic practices;
4. *Political organization.* The political structure of the host country and of particular resorts is important. Such factors as the existence of capitalist or socialist principles; planning regulations, incentives and constraints; and the roles of national, regional and local tourist organizations influence tourist impacts;
5. *Level of tourist development.* This encompasses the degree of local involvement in the tourist industry; rate of development; nature and diversity of attractions; types and quality of accommodation; entertainment and eating facilities; and the role of travel intermediaries.

The capacity of a destination to absorb the demands imposed upon it by tourists and the tourist industry depends upon the interrelationships of the numerous, complex factors itemized above.

Discussion

Capacity levels are influenced both by the characteristics of tourists and of the destination area and its people. For example, host irritations caused by the presence of tourists may be intensified in cases of concentrated use, such as package tours, by long periods of stay, by the continued demonstration of the material superiority of visitors when compared with permanent residents, and by the selection of activities which bring the host and guest into close contact. Host resentment is likely to be high in locations with a highly developed tourist industry but with only limited local involvement.

Given the multiplicity of factors which influence the nature of capacities, it is difficult, if not impossible, to calculate absolute values for tourist carrying capacities. However, carrying capacity has implications beyond the mere quantification of capacity levels. Certain sub-systems may have low tolerance levels to tourist activity while others may be exceedingly tolerant. Identification of these, and knowledge of the consequences of exceeding tolerance levels, could lead to the implementation of preventative management controls and maintenance policies. Destination areas have limitations in the volume and intensity of tourist development which can be absorbed before deterioration becomes a major problem. In the chapters which follow it will be seen that some tourist areas have been unable to absorb tourist demands and capacities have been exceeded. If this occurs, the very resources which attracted the tourists initially can be destroyed.

The consequential element

The concentration of tourists and associated facilities and services in destination areas gives rise to a variety of impacts. Economic impacts encompass the monetary costs and benefits which result from the development and use of tourist facilities and services. Physical impacts include alterations to the natural environment, including air, water, soils, vegetation and wildlife, as well as changes in the built environment (Wall and Wright 1977). Social impacts are the changes in the way of life of residents of destination areas. Carrying capacity levels differ between the economic, physical and social sub-systems of destination areas and also vary among the components which contribute to any one of these three sub-systems. Therefore a tourist activity can be economically desirable while also being socially and environmentally damaging. For example, a tourist activity may create jobs and income for residents but working hours may be such that family life, social behaviour and recreation are adversely affected. Tourist developments may be environmentally damaging but may also bring in considerable revenue which, in turn, can be used to ameliorate pollution. The links betweeen each category are so close that the measurement of and planning for tourist impacts is rendered complex and difficult.

The development of measures which will mitigate the negative impacts of tourism and, at the same time, enhance its positive effects will undoubtedly involve trade-offs. In reaching such decisions, knowledge of the dynamics of tourist impacts and how these vary with different levels of use, different tourist activities, and tourist areas with different characteristics is imperative. Such information is not yet available and, as a result, planning for tourist impacts has been rudimentary.

Recognition of the need to plan for or manage tourist impacts has arisen largely from the existence of negative or crisis situations, and has often only occurred when carrying capacities have been greatly exceeded. In some cases it has been necessary to restrict the use of tourist facilities and barriers have been erected preventing access to points of interest. Stonehenge is a good example of this situation. The stones were beginning to suffer from visible wear and tear. This was not due primarily to wilful damage but, rather, was the result of the cumulative effects of thousands of feet trampling the foundations and thousands of hands touching the stones, particularly the shallow carvings. Faced with the dilemma brought about by the success of Stonehenge as a tourist attraction, it was decided that preservation was more important than complete freedom of access and, in May 1978, fencing was erected and a new system of viewing was introduced. Visitors are no longer allowed within the stone circle. They are free to wander around the outside of a perimeter fence and must keep to a pathway if they wish to get a closer look at the stones (Bainbridge 1979).

The range of measures for modifying the impacts of tourism is large. Finance may be made available to rehabilitate historic sites and it is also necessary to repair more modern facilities suffering from the effects of over-use. Zoning regulations, building codes and design standards may be employed to ensure that new structures, such as hotels, are constructed in appropriate locations and are of acceptable appearance. However, to date such measures have usually been employed after the damage has been done. They have often been site-specific rather than reflections of an imaginative, general policy. They have usually been reactionary, discouraging unsightly developments, rather than positive, promoting excellence in design.

The organizational framework which has been presented in Fig. 1 is an attempt to summarize some aspects of the dynamics of tourism. It stresses linkages between tourist use, carrying capacities, and the impacts of tourism. It recognizes that destination areas have tolerance limits and suggests, by implication, the replacement of reactionary planning and management measures by preventative planning strategies.

Tourist decision-making

The conceptual framework which has been discussed addresses some of the more important factors and relationships influencing impacts of tourism. These impacts, ultimately, stem from the multitude of individual decisions to visit particular places and to participate in specific activities. It is appropriate, therefore, to give some attention to the decision-making processes of travellers. This aspect of travel research has been little studied and, as a result, is not widely understood.

It has been suggested that carrying capacity can be manipulated as a means of controlling the effects of tourist development. Our consideration of a decision-making framework implies that part of the planning process can be directed at tourists and the factors which govern their travel decisions. An understanding of decision-making processes may be used as an aid in planning in four ways (Schmoll 1977: 60):

1. Through marketing. Through promotional schemes, tourist decisions may be directed in favour of particular destinations, accommodation types, modes of transportation and travel services. Traffic may be diverted away from areas which have reached saturation point by encouraging potential tourists to select alternative destinations;
2. Through the identification of the factors which have a bearing on travel decisions. This should indicate which positive influences should be reinforced and which undesirable influences should be counteracted;
3. Through the identification of areas requiring more research. Improved knowledge should lead to more informed planning decisions;
4. Through the determination of criteria by which target markets, or market segments of special interest and value to a tourist enterprise or destination, can be identified.

Preferences for destinations change. Accurate forecasting of travel patterns requires an understanding of the features of destinations which appeal and stimulate travel to them. By manipulating the factors influencing tourist travel decisions, planning can be more effective and preventative. Thus the costs of mitigating the effects of tourist development can be reduced or avoided.

The vacation travel market has become highly competitive. Increased discretionary time and money have given the potential tourist flexibility in choice. Tourist destinations have responded by becoming more competitive in both the prices and qualities of the facilities and services which they provide. As a result the factors influencing tourist decisions have become increasingly more complex.

Early approaches to the study of decision-making were based on the concept of economic man. This concept assumed perfect knowledge and economic rationality in decision-making. In the case of tourism, this would mean that tourists would arrange themselves in space and time to optimize the benefits of travel within the constraints of the amounts of time and money available. However, tourist travel involves elements of uncertainty. For instance, the climatic conditions of destinations may not be fully appreciated and weather may be variable. The quality of services and facilities may be different from what is expected. Furthermore, tourist experiences are constrained by the characteristics of the destination and these may not be fully appreciated by visitors, particularly if they are visiting for the first time.

The deficiencies of the normative, economic-man, approach stimulated its replacement by a behavioural, or satisficing, perspective (Simon 1957). This assumes that tourists, still acting rationally but on the basis of limited information, seek satisfactory rather than optimal experiences. Behavioural decision-making models reflect this different emphasis and, according to Rostron (1972: 38), are more applicable in understanding how people reach decisions and act upon them. A behavioural decision-making framework has been presented for travel and tourism in an attempt to identify, understand and illustrate the factors influencing tourist decisions and behaviour (Fig. 2). Impacts of tourism result from behavioural outcomes of the tourist decision-making process.

If viewed in economic terms, the tourist is engaged in a buying decision, spending money to gain satisfaction. In typical consumer product purchases, the buyer usually expects predominantly tangible returns, the decision is largely spontaneous and involves only a small part of the consumer's assets (Wahab, Crampon and Rothfield 1976: 74). The buying decisions of tourists are unusual in several ways:

1. There is virtually no tangible return on the investment. Exceptions to this statement arise from business travel where contracts may be signed and financial transactions may take place. Tourists purchase souvenirs and gifts on their trips but these usually represent only a small proportion of total expenses. The tourist product is an experience rather than a good;
2. Expenditure is often substantial. The purchase of a tourist package involves much larger monetary outlays then most other consumer purchases;
3. Purchases are not usually spontaneous. Trips are normally carefully planned, particularly in terms of expenditure. Small purchases, such as

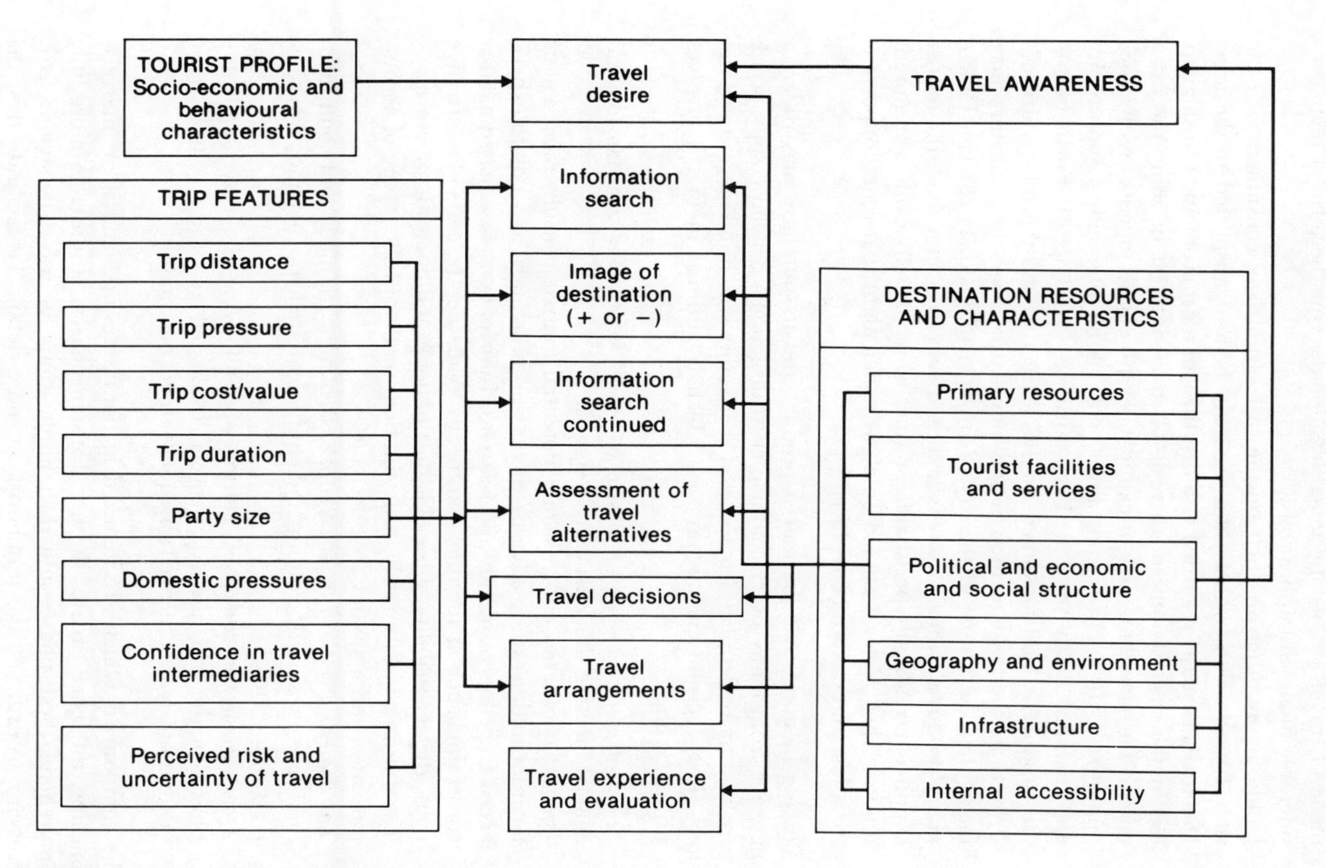

Fig. 2 The tourist decision-making process

gifts and souvenirs, may often be spontaneous but these usually have relatively little cost. Choices of destination, type of accommodation and mode of travel are usually much less spontaneous because of their considerable financial implications;

4. Unlike many other products, in the case of tourism, consumers visit the site of production, rather than the good being transported to the user. Distance is usually regarded as a disutility in most economic transactions but in the case of tourism there is often an interest in going the extra mile to see something new or exotic. Clearly, not all tourists are distance minimizers. In such circumstances, notions of intervening opportunity and distance decay may need qualification (Wall 1978). Furthermore, the product is not consumed in the sense that most products are destroyed in their use, although the cumulative impacts of visitors may mean that the nature of the tourist product may slowly change. Unlike many products, which can be stored and sold at a later date, the tourist supplier has a highly perishable product which usually cannot be stored: if a hotel bed is not occupied on a particular night that rental opportunity is lost and cannot be replaced.

Bearing in mind that the tourist product and its purchase may differ in some ways from many other commercial transactions, it is now appropriate to turn to a more detailed discussion of the factors which influence travel decisions. The decision-making process, which is outlined in Fig. 2, involves five principal phases:

1. *Felt need or travel desire.* A desire to travel is felt and reasons for and against meeting that desire are weighed;
2. *Information collection and evaluation.* Potential tourists consult travel agents for information, study advertisements and brochures, and talk to experienced travellers or friends and relatives. Information is evaluated against constraints such as the time and money available, the costs of alternative trips, the accessibility of possible destinations, and domestic pressures;
3. *Travel decisions.* Ultimately, a destination, mode of travel, type of accommodation and activities are selected;
4. *Travel preparations and travel experience.* Bookings are made and confirmed, funds are organized, clothing and equipment are arranged and, eventually, travel begins;
5. *Travel satisfaction evaluation.* During and following the travel and stay phases of the trip, the experience is evaluated and the results of these evaluations will influence subsequent travel decisions.

The components which are itemized in the framework, and their interrelationships, influence each of the five decision phases. In order to more fully comprehend these phases and their implications for travel behaviour, it is necessary to examine the framework in more detail. To facilitate this, the framework will be considered under four major headings: the tourist profile, travel awareness, trip features, and the resources and characteristics of destinations.

The tourist profile

The tourist profile can be viewed under two major categories: the tourists' socio-economic and behavioural characteristics.

Socio-economic characteristics of tourists

Age, education, income and previous experiences influence attitudes, perceptions and motivations and affect decisions. Obviously, individuals with low disposable incomes are less likely to pursue travel arrangements which involve first-class air fares, expensive hotels, and costly restaurants, than those who are more wealthy. Few elderly travellers are likely to participate in mountain climbing or visit night-clubs. Relationships between socio-economic variables and participation in recreation and travel are well documented in the literature and no further consideration of them will be given here. However, it should be noted that such relationships do not really explain recreational behaviour: it is not possible to predict what people will do merely from knowledge of their socio-economic characteristics.

Behavioural characteristics of tourists

The motivations, attitudes, needs and values of tourists are of crucial importance in contributing to their decision-making processes. Motivations to travel are related to expectations, needs and wants. These, in turn, reflect tourists' personalities and socio-economic profiles. Although highly interrelated, for the sake of simplicity, these variables will be discussed separately under the four following headings: needs, motivations, travel awareness and attitudes.

Needs.

The need of an individual is an internal condition that arises from the lack of something which, if present, would tend to further the well-being of that individual. Needs may be emotional, spiritual or physical. Collectively, they are determinants of tourist motivation. An individual's need for change, new experiences, adventure and aesthetic appreciation may all be satisfied by travel and tourist activity.

Motivations.

The question of why people travel has been given little attention until the recent research of Lundberg (1971), Mercer (1976a, b), and McIntosh (1977). Wahab (1975: 44) described travel motivations as: 'one of the most basic and indispensable subjects in tourism studies. Ignoring it or passing lightly over it would defeat the whole purpose behind tourist development plans.' In early times, people travelled for reasons which were more easily identified than the reasons inducing modern tourists to travel. These included the search for food, water and shelter, for economic and territorial gain and religious devotion. Few people travelled for sheer pleasure. Modern tourists travel essentially for pleasure but may do so for a number of additional reasons. These consist of push factors, such as the desire to escape the daily living and working environment and the pollution and traffic congestion of cities, and pull factors, such as the attractions of the destination, the visiting of friends and relatives, or to play or watch sport.

Table 1 Motivations of the tourist

Motivational category	Motivations
Physical motivations	refreshment of body and mind; for health purposes (i.e., either medically prescribed or undertaken voluntarily); for participation in sports; pleasure – fun, excitement, romance and entertainment, to shop.
Cultural motivations	curiosity about foreign countries, people and places; interests in art, music, architecture, folklore; interest in historical places (remains, monuments, churches); experiencing specific international and national events e.g. Olympic Games, Oktoberfest.
Personal motivations	visiting relatives and friends; meeting new people and seeking new friendships; seeking new and different experiences in different environments; escaping from one's own permanent social environment (i.e. desire for a change); personal excitement of travelling; visiting places and people for spiritual reasons (i.e. pilgrimages); travelling for travel's sake.
Prestige and status motivations	pursuit of hobbies; continuation of education or learning; seeking of business contacts and professional goals; conferences and meetings; ego enhancement and sensual indulgence; fashion i.e., 'keeping up with the Jones's'.

Note: Some of the above motivations cannot be distinguished empirically from other groups and are not articulated by tourists. For instance, it is readily known that travelling is fashionable and is pursued because of the 'snob value' associations it often entails.
Source: After McIntosh 1977

A number of researchers have attempted to classify tourist motivations (Thomas 1964; Gray 1970; McIntosh 1977). However, there is little agreement concerning the relative significance of each motivational component. Thomas (1964: 65) listed eighteen factors and Lundberg (1972: 128) listed twenty. Both inventories drew heavily upon the attributes of tourists' personalities and personal value systems. Gray's (1970) categorization was far less explicit. He dichotomized travel motivations to be part of either 'sunlust' or 'wanderlust'. The former referred to the seeking of locations containing better or different amenities to those available at home. 'Wanderlust' referred to the experience of seeing new places, meeting new people and engaging in new and different activities.

Tourist motivations are far more complex than these classifications suggest. Motivations operate at a general level when they induce a person to travel. They become more specific when a tourist is urged to visit certain destinations or select a particular activity or mode of transport. Common tourist motivations are listed in Table 1. Tourist motivations are diverse and may be incompatible. Motivations of travellers visiting places on business, for health reasons, or to see relatives may be largely predetermined. In such circumstances, choice of destination may be beyond the control of the traveller and commitment to the destination may be weak. This has obvious implications for the likelihood of return visits.

Travel awareness.

Potential tourists may be motivated to travel but, unless they are informed of what opportunities are available, they may be unaware of the means of meeting their requirements. Awareness of destinations, facilities and services depends upon the availability of information and the credibility of its source. Information on the tourist product is transmitted to potential tourists through both formal and informal sources. Formal sources include magazines, travel brochures, advertisements on radio and television, and discussions with travel intermediaries. Informal sources refer to comments obtained from relatives, friends or other travellers. Nolan (1976: 7) noted that travel information received from relatives and friends was the most informative, but it ranked lowest in credibility. Information obtained from guide books, government tourist offices and automobile club trip planning services ranked highly both in terms of quality and credibility.

A tourist image, or impression, is conjured up from the information received, as interpreted through the personal and behavioural characteristics of the tourist. Lawson and Baud-Bovy (1977:10) defined image as: 'an expression of all objective knowledge, impressions, prejudices, imaginations and emotional thoughts an individual or group have of a particular object or place'.

The decision to travel precipitates a series of subsequent decisions, including choice of destination, mode of travel, length of stay and type of accommodation. The images of potential destinations are important to these decisions. The quality and range of services may be similar in a number of destinations but their images may differ. These differences may be the decisive factors in the choice between travel opportunities. The images of destinations may differ greatly from reality. The larger the difference between image and reality, that is between expectations and experience, the more likely is the tourist to be dissatisfied. Therefore, information made available to the tourist should have a firm basis in reality.

The effects of image on tourist decision-making have received little research attention although equivalent studies have been undertaken for the marketing of other consumer products. Some studies have attempted to describe tourist images (Ehemann 1977) but few, with the exception of Hunt (1975), have related them to decision-making. The role of images in influencing tourist product preferences and, ultimately, decisions is, therefore, a potentially fruitful area of research.

A number of possible destinations may appear to satisfy the personal requirements of the potential traveller. Each alternative is evaluated according to its compatibility with a number of criteria which are indicated under the three broad headings of generation point characteristics, trip features, and destination resources:

1. *Generation point characteristics*. Attributes of the place of residence, or generation point characteristics, refer to considerations regarding home, work and family. They might encompass willingness to leave work unfinished around the house, the responsibilities of taking children travelling, selecting destinations which all members of the family will enjoy, making arrangements for the care of family pets, for job replacements, and for the welfare of family members in the absence of the traveller. Travellers may not select destinations at great distances because of these deterrents, nor will they stay very long if their jobs demand their services. Trips are less likely to be multi-destinational if young children are to be included in the party;
2. *Trip features*. These include such factors as distance, duration of stay in one or more destinations, time constraints, trip cost and value for price, party size, perceived risk and uncertainty at the destination, and confidence in travel arrangements and travel intermediaries;
3. *Destination resources*. The features of potential destinations play an important role in the assessment of alternatives and in the final choice of destination. These include types of attraction, the availability and quality of services, environmental conditions, the attributes of the host population and their political organization. In addition, the potential tourist is likely to consider the practical barriers of entering a destination (customs, immigration and security inspections), the accessibility of the destination and of other complementary attractions, and host attitudes towards tourists. As most of these features were considered in the discussion of the conceptual framework of tourism (Fig. 1), there is no need to examine them in detail here.

Attitudes.

Tourist attitudes provide indications of the attractions of travelling and visiting distant places (Murphy 1975: 216). Attitudes are important components in travel motivation. They reflect past experience (Neulinger and Breit

1971: 108). For example, a visit to Rome to participate in the Easter celebrations at the Vatican may have been marred by large crowds, the commercialization of what the visitor hoped would be a meaningful spiritual occasion, and by language difficulties. The experience may have induced a negative attitude towards subsequent trips to Rome, particularly at Easter, and to religious festivals in general. This may deter the traveller from participating in similar trips in the future.

Information on many of the above concerns will be revealed to the potential tourist via the formal and informal information sources mentioned previously. Knowledge of some aspects will have been gained from earlier travel experiences, and some considerations may be unperceived.

The combination of trip features and destination resources constitute the information base upon which decisions concerning destinations, modes of travel and activities are formulated. Generation point characteristics place constraints upon these decisions. (For ease of presentation, trip features and generation point characteristics have been combined under the former heading in Fig. 2). Following these decisions, final travel arrangements can be made. After returning from the trip, tourists recollect and evaluate their total travel experiences. These experiences provide the bases for subsequent travel decisions, whether to the same destination or another. These decisions, and the resulting behaviour, give rise to impacts on destination areas which are the concern of much of the remainder of this book.

Spatial relationships

Tourism, by definition, involves the movement of people from their places of permanent residence to new locations. This means that decisions made in one location have implications for other places which may be far removed from the locations in which the decisions are made. The preceding discussion has concentrated upon the decisions made by tourists. However, numerous other people also make decisions concerning the deployment of tourist resources. In addition to the individual tourist who has to decide when and where to participate, and what forms participation should take, there are numerous agencies with a responsibility for monitoring, regulating and providing tourism and recreational opportunities. Other groups, such as travel intermediaries and advertising agencies, influence decisions. Most major investment decisions are made by governments and multi-national corporations. The homes of most tourists, the head offices of most hotel chains and transportation companies are in the major cities of the developed world, and this is where the majority of crucial decisions concerning tourism are made.

The concentration of decision-making in the large cities of the developed world has caused some authors to view tourism, both international and domestic, as involving metropolitan demands being met by peripheral supply and to suggest that the core–periphery concept could be of utility in tourist studies (Christaller 1963; Britton 1980). It has also prompted charges of economic imperialism as the major decisions affecting the economies of destination areas are made in places beyond their control. These charges will be examined in more detail later.

Not all tourism is of the centre–periphery type. In fact the major cities of the world constitute principal sources of tourists and are important destinations in their own right. Such cities as Paris, London and New York are among the world's most popular tourist destinations. Most big cities are multifaceted tourist attractions. They possess a wide range of facilities, including museums, art galleries, theatres, cinemas, restaurants, specialized shopping facilities, public buildings, sports stadiums and parks, which tend to be clustered in relatively small areas. Many of these are high order functions which require large numbers of people for their sustenance, and which have difficulty surviving in areas of low population density. Large cities are also centres of communication. They are the foci of road, rail, and air transport networks which channel travellers in their directions, and the transportation and accommodation infrastructure which is required for the functioning of any city is also necessary for tourism. Tourism does not dominate the economies of big cities as it does in resort areas, but hotels are an increasingly important component of big city landscapes and are a sign of their significance for tourism (Vetter 1975; Hutchinson 1980; Wall and Sinnott 1980).

Summary

The two frameworks indicate that impacts of tourism are the outcomes of the interrelationships of a complex array of phenomena. The first framework is much more general than the second. It shows that impacts of tourism are the result of interactions between tourists and the destination area and its population. The economic, physical and social sub-systems of the destination area have carrying capacities and the magnitude and direction of tourist impact is determined by the tolerance limits of each. Positive impacts are present until limits are exceeded when impacts become negative.

The second framework is more specific than the first and focusses upon one important component of the former: the tourists and their decision-making processes. The decision-making process outlined in Fig. 2 indicates that impacts of tourism are the consequences of tourist decisions. It recognizes that impacts of tourism are dynamic, changing with corresponding changes in destination features, trip characteristics, and the personal and behavioural attributes of tourists. There is a spatial discontinuity between cause and effect, as most of the decisions are made in the urban areas of the developed world, and many of the consequences are felt in the pleasure periphery.

The next three chapters, on economic, physical and social impacts, constitute a detailed examination of the consequences of tourist decisions on each of these sub-systems in destination areas.

3 economic impacts

Foreign visitors to Canada spent approximately $10 billion (Canadian) in 1977. This was approximately 5 per cent of the Gross National Product. The Canadian Government Office of Tourism estimated that, directly and indirectly, these expenditures created 900,000 jobs for Canadians. These results were achieved with an annual investment of $1.5 billion (Canadian), from public and private sources, to meet the need for services generated by travel to and within Canada. In countries such as Jamaica, Spain and Mexico, tourism is the largest earner of foreign exchange and the leading industry in terms of income and employment. In 1978, for the first time, tourism emerged as the top foreign currency earner in Britain. The 11.5 million foreign visitors spent £2,750 million in Britain in the Jubilee year of 1977, pushing takings up by 14 per cent on the previous year, and extending tourist employment figures to 1.5 million jobs. These international tourism statistics only partially indicate the economic significance of the industry. More tourists travelled and spent even larger amounts of money within their own countries. In Great Britain alone, for example, over 3.5 million domestic vacations have been recorded annually since 1971, with expenditures totalling more than £1,500 million each year. The economic contributions of domestic tourism are as significant as those of international tourism but there is a paucity of reliable statistics on expenditures and, therefore, it is difficult to define its impact accurately.

There is no doubt that tourism has major effects on the economies of destination areas. Research has focussed, primarily, upon the economic aspects of the industry and this emphasis has resulted in a proportionately large number of studies of these effects. The majority of studies of the economic impacts of tourism have been directed at international and national levels (Peters 1969; Gray 1970; Thuens 1976), with fewer investigations at regional and local levels (Archer, Shea and Vane 1974; Henderson 1975; Vaughan 1977a, b; Wall and Knapper 1981). Although economic studies, such as Ogilvie's (1933), were among the earliest reports on leisure and recreation, when compared with the latter most economic studies are of comparatively recent origin. In spite of recent increases, studies of tourism, particularly in North America, constitute only a small proportion of works on the broader topics of leisure and recreation. The early studies of Ogilvie (1933), Alexander (1953) and Waugh (1962) offered introductory statements on the economics of tourism but they did not provide a detailed examination of the full array of economic effects which have been mentioned in recent publications.

Most studies have emphasized the economic benefits which accrue to destination areas. The development of tourist facilities and recreational opportunities has frequently been viewed as a major positive contribution to the national balance of payments, and as a means of redressing regional disparities in incomes and employment. Until recently only a few studies examined the economic costs of tourism, and the direct costs of entering the tourist market, as well as the indirect costs to destination areas, received little attention (Young 1973; Jafari 1974; Turner 1976). In spite of the uneven emphases within economic studies of tourism, a concentration upon economic questions has occurred at the expense of research on physical and social impacts of tourist developments.

A number of factors have contributed to the economic emphasis of a majority of tourist impact studies. Firstly, when compared with physical and social impacts, economic impacts are relatively easy to measure. Physical and social impacts, particularly the latter, are difficult to subject to numerical analyses, for they are often composed of intangibles and incommensurables which are difficult to quantify. Few researchers have accepted the challenges posed by the qualitative aspects of tourist impacts. Furthermore, there are widely accepted methodologies for measuring economic impacts, but they are still in the early stages of development in the other social and environmental sciences.

Secondly, large quantities of relatively reliable and comparable data have been collected on economic aspects of tourism. Data required to measure the economic costs and benefits of tourism, such as expenditures, employment and tax revenues, have been more readily available than for other tourism impacts. Many of these data have been collected routinely by government agencies.

Thirdly, the emphasis on the economics of tourism, especially its benefits, reflects the widespread belief among agency personnel that tourism can yield rapid and considerable returns on investments and be a positive force in remedying economic problems. Governments, development agencies, financial organizations, planning departments, local councils and other groups that support and promote the tourist industry have often seen tourism as a means of counteracting the economic difficulties that they have been facing. It is not surprising, then, to find that most research on the economic benefits of tourism has been conducted, instigated or sponsored by such institutions. In Canada, at the federal level, such studies are undertaken by the Canadian Government Office of Tourism. Provincial governments also produce economically-oriented publications through such departments as the Ontario Ministry of Industry and Tourism. In the United Kingdom a large proportion of economic studies of tourism bears the names of the British Tourist Authority, which is primarily concerned with international travel, and the English, Scottish and Welsh Tourist Boards, which initiate and sponsor research at a regional level. Numerous county authorities, including Gwynedd, Cumberland, Devon and Northumberland, have also undertaken economic studies of tourism (Archer 1973: 1).

With the advent of a number of independent international agencies, economic consultancy firms and university departments, a broader view of tourist economics has emerged. Research emanating from the Tourism and Recreation Research Unit of the University of Edinburgh, from the Department of Economics at the University College of North Wales, from the Department of Habitational Resources, University of Wisconsin (Stout), from consultancy firms such as A. D. Little Inc. and D'Amore and Associates, and from the Economic Commission for Europe has been more penetrating and comprehensive. The establishment of associations to collect, organize, standardize and publish travel data has also increased the amount and quality of recent economic research in tourism. One example of this type of association is the Discover America Travel Organization (DATO). Waters (1974: 2) believed that the co-ordinating function of this body would improve economic research on tourism. The US Travel Data Center is a subsidiary of DATO. It aids in the planning of American tourism by increasing the quality and availability of data. The International Union of Official Tourist Organizations has also had a marked influence on the collection and dissemination of information on tourism.

As a result of such administrative and academic developments, research on the economics of tourism has moved beyond the documentation and description of economic benefits as indicated in tables of international travel statistics. Recent progress includes:

1. The measurement of secondary economic impacts;
2. The improvement of techniques for the analysis of travel data;
3. The initiation of research on the economic costs of tourism.

This chapter examines the economic impacts of tourism as indicated by the current literature. The first section describes some of the economic characteristics of tourist development. It is followed by a consideration of the role of tourism in promoting economic development. More specific economic effects of tourism are the subject of the third section which is divided into economic benefits and economic costs. The chapter is concluded with a summary. Inevitably, there are overlaps between sections and conflicting statements exist in the literature. Many authors reveal underlying optimistic attitudes towards tourism, particularly among government spokesmen in less developed countries, while others indicate that some governments are beginning to reassess the role which tourism might play in their national economic development. Even though separate subsections are presented for economic benefits and economic costs, it will sometimes be appropriate to discuss costs alongside benefits to provide a more balanced assessment of the study of economic impact.

Economic characteristics of the tourist industry

Tourism is often welcomed as an industry bringing desperately needed foreign exchange, employment, and a modern way of life. To others tourism raises the spectre of the destruction of traditional lifestyles, neo-colonialist relationships of exploitation, overdependence upon an unreliable, single industry and inflation (Schneider 1976: 5). Examples in the literature support the arguments of both camps. Although, because of the heterogeneous nature of its facilities and services, some have questioned whether tourism should be regarded as a single industry, there are a number of unusual and unifying particulars of tourism which make it distinguishable from other industries and international transactions (Sadler and Archer 1974: 2–3; Schneider 1976: 9–10; Leiper 1979: 397–403).

Firstly, tourism is an invisible export industry. As in the cases of banking and insurance, there is no tangible product which is shipped from one place to another. It is one of the few industries in which the consumer actually collects the service personally from the place where it is produced. In consequence, the exporting destination incurs no direct freight costs outside its boundaries except where the transportation facility used by the tourist is owned by the destination. In developing countries this is seldom the case and tourism is sold free on board (f.o.b.). In developed countries, airlines are frequently owned by the exporter and hence tourism is sold both f.o.b. and c.i.f. (cost, insurance and freight).

Secondly, tourists visiting destination areas require ancillary goods and services, such as transportation facilities, water supplies, sewerage systems and retail functions. These have to be created, expanded or imported, depending upon the availability of existing supplies and the nature and magnitude of the tourist demands.

Thirdly, tourism is a fragmented product, integrated with and directly affecting many other sectors of the economy. Tourists use and consume a spectrum of components, some of which are purchased from firms specializing in tourist businesses, some from firms in other industries, and some are derived at no direct cost to the tourist. For example, tourist expenditure is injected directly into hotels, shops, restaurants and recreational facilities. Indirect benefits from tourist expenditure may be in the form of local tax revenues, improvements in the infrastructure of destination areas and extensions of community services. Demands by tourists for specific items, such as souvenirs, stimulate local entrepreneurial activity, providing additional local employment and income. The large number of forces at play within an economy make the full measurement of economic impacts a highly complex undertaking.

Fourthly, tourism is a highly unstable export. It is subject to strong seasonal variations, to pronounced and unpredictable influences from external forces, to the heterogeneous nature of tourist motivations and expectations, and is highly elastic with respect to both price and income. Collectively, these factors promote a low level of customer loyalty with respect to destinations, modes of travel, accommodation units and travel intermediaries (Schmoll 1977). These factors are examined in more detail below:

1. The fact that the tourist product cannot be stored, and that tourist demand is highly seasonal lead to marked fluctuations in levels of activity in the industry. This means that sufficient must be earned during the peak season to offset a decline in patronage for the remainder of the year. The cyclical pattern of demand for tourist goods and services has obvious implications for employment and investment.

A recent and growing trend has been for tourists from developed nations to take two or more holidays in one year although the secondary vacations are often spent in domestic locations. Nevertheless, this has mitigated, to some extent, the seasonal peaks in such tourist destinations as Mexico, Majorca, Morocco, the Caribbean and the Pacific (World Bank 1972: 9). Planned measures to offset the problems created by seasonality may take one of two approaches:

(a) Altering the rate of production of supply to correspond more closely with the peaks in tourist demand. Suppliers have two alternatives in selecting this option. They may attempt to meet peak demand at the expense of reducing the qualities of the services provided, or they may restrict supply at a level below the peak demand. The latter option may cause some tourist dissatisfaction but it ensures that tourists whose demands are met do not receive a diminished quality of services. Extensions of supply during peak periods stimulates further temporal concentration of demands leading to increased overloading at existing facilities;

(b) Modifying the temporal distribution of demand to match existing levels of supply. There have been few successful attempts at altering the temporal incidence of demand. Staggering school holidays has achieved little in this respect. Off-season concession rates offered by airlines and hotels, and the off-season staging of business conventions have been more fruitful in extending the season. However, the complete success of these measures has been questioned because of the limits of price elasticity: price cuts cannot be very large if profitability is to be maintained. The effects of seasonality probably cannot be totally removed and seasonality is a factor with which the tourist industry must learn to live;

2. Tourism demand is subject to change from unpredictable external influences. Long-distance pleasure travel is a luxury. Political unrest at particular destinations, changes in international currency exchange rates, energy shortages, and unusual climatic events can cause tourist traffic to divert to new destinations with more amenable conditions;
3. Motivations of tourists are highly complex, often incompatible, and vary greatly among travellers. As a result, many tourists seldom travel to a particular distant location more than once. Destinations have to assess which segments of the tourist market they are in the best position to satisfy, given the types and qualities of tourist product which they can create. The creation of product loyalty and the attraction of return visitors is particularly difficult when the lure of the unknown and exotic is beckoning;
4. Tourism is price and income elastic. This means that tourist decisions will be greatly influenced by small changes in price and income. Price elasticity is more easily identified than income elasticity because of its frequent and sudden effects. Income changes are usually more gradual and their effects on demand may be identified only after a period of years. Airline price wars between North America and Europe, and North America and Australasia, have had significant effects on the volume of tourists travelling to these destinations. The Laker Skytrain between

London and New York was an excellent example of price elasticity, for relatively low fares generated new traffic. However, evidence concerning the significance of price elasticity is mixed (Archer 1975: 22). Quoting a study by Bond and Ladman, Archer concluded that near unity (i.e. price inelastic) figures for American travel to Latin America and to the West Indies were because alternative destinations were still more expensive in 1972. Thus little travel was diverted by a change in price structure. This might be less true of the current price situation *vis-à-vis* air-fares from the United States.

The volume of tourist traffic can be manipulated by international price and currency controls. By restricting the volume of currency which can be taken abroad, and by means of import duties and export taxes, countries of tourist origin can create exchange rates for tourists which differ from those at which other foreign trade takes place. Somewhat similar subsidies, as in the case of duty-free concessions, can be made by destination areas which effectively permit tourists to purchase goods at less than indigenous prices.

The characteristics of the tourist industry which have been outlined above serve as a partial foundation for understanding the nature of tourist impacts. They also go some way towards explaining differences between the impacts of tourism and some other forms of recreation. They also alert the reader to a number of basic concerns in the development of tourism which are often overlooked by planners but which are of considerable significance in the assessment of the costs and benefits of tourism.

Tourism and economic development

Tourist development gives rise to different benefits and costs in different areas. Many of these differences are attributable to variations in the economic structures of destination areas and their geographical locations. The most obvious distinction is that between developed and developing areas. Developing countries usually have low levels of income, uneven distribution of income and wealth, high levels of unemployment and underemployment, low levels of industrial development which are hampered by the small size of the domestic market, a heavy dependence on agriculture for export earnings, and high levels of foreign ownership of manufacturing and service industries. These trends have been associated with large regional disparities in economic wealth within many of the developing countries, a substantial leakage of profits out of the country, high inflation and shortages of foreign exchange. Although many of these economic problems also plague developed countries, they have usually been caused by different economic factors. The rapid injection of tourist expenditures and foreign investments into developing countries often has rather different and more significant effects than if equivalent sums were expended in developed economies. The directions of monetary flows, the distribution of benefits, employment characteristics and income effects will vary greatly with the sources of tourists and investments, and with the nature and level of economic development of the destination.

Tourism in developing countries is, in most cases, a relatively new activity which has grown to significant levels over a very short period of time. This has resulted in a heavy strain being placed upon local infrastructures and human resources. In some cases, the infrastructure is not adequate to absorb the increase in tourist arrivals. In other situations, as in the case of India where a number of first-class hotels have been built, the relatively small absolute number of tourists may be insufficient to fully utilize the available facilities. In the latter circumstance it may be necessary to increase prices to offset the costs of underutilization. New tourist activity may be imposed upon traditional cultures with different standards of living and socio-cultural backgrounds from those of the tourists with whom they interact. This is, in itself, responsible for numerous economic, physical and social impacts peculiar to the tourist industry.

A considerable body of literature emphasizes export expansion as a major factor contributing to sustained economic growth. Within developing countries, it has been frequently argued that a transformation from a traditional agricultural economy to an industrial economy is required for modernization and economic development to take place. Such a transformation would demand enormous amounts of capital and foreign earnings or loans. However, many developing countries have a strong historical function of exporting primary commodities and the export of primary products has usually been insufficient to meet the financial requirements of the proposed economic transformation. This has encouraged governments in the developing world to turn to tourism as a means of acquiring the financial resources required for industrialization. Somewhat similarly, at a more local level, tourism has been viewed as a means of introducing new growth into declining rural economies in developed countries (Brownrigg and Greig 1976). Peters (1969: 10) has summarized the views of the proponents of tourism as an agent of economic development as follows:

> The economic gap between rich and poor countries has widened over the past ten years. But to create new industries and to transform rural life in Asian, African and Latin American countries is a gigantic task. The relevance of tourism to this situation is that income from international travel can bring the foreign exchange essential for major investment. There is a widespread awareness of the potential benefits, but little has been done in practice to provide the means of expansion of tourism plans in most of the developing areas of the world.

In addition to general arguments concerning the ability of tourism to produce foreign exchange, advocates of tourism have also been more specific in their claims. Proponents of tourism development in developing countries have argued that, not only can tourism relieve the shortages of foreign earnings constraining economic development, but it can also alleviate problems of urban unemployment and, in the long term, provide a price and income elastic substitute for traditional exports which face less secure futures (Diamond 1977: 539). As a result of such arguments, the case for the promotion of tourism as a strategy for economic growth has received widespread approval, particularly among policy-makers in developing countries.

Tourism, however, has not escaped criticism. Indeed, the challenge to the industry is a mounting one, growing continually in volume and insistence. Bryden (1973), Economist Intelligence Unit (1973), Perez (1974), Rivers (1974), Marsh (1975a, b) and Turner (1976) are among those who have expressed reservations concerning the benefits of tourism. Most criticisms have not been concerned with the economic potential of tourism, an argument which is largely unassailed, but rather with the negative non-pecuniary effects of tourism. Accompanying the widespread economic benefits, there are a variety of unquantified physical and social costs which may be of sufficient magnitude to support arguments against its further expansion (Bryden 1973: 1). These criticisms have made little impact upon governments and planners, as tourism is still viewed by many as an important component in their plans for economic development. This chapter is not concerned with the nature of these arguments as they are considered later in the book.

A large proportion of studies which examine the significance of tourism for developing countries have attempted to isolate ways in which tourism can contribute to the process of economic development. They include supplements to the national balance of payments, the creation of employment, the nature of infrastructural investments and the external economies created, intersectoral linkages, and the multiplier effects of tourist expenditures. Taken together, these studies offer powerful support to those arguments which encourage countries to promote their tourist industries. However, few studies have explicitly examined the underlying rationale of tourism as a means of economic development, the successes and failures of countries using tourism for that end, and the problems and precautions involved in planning for tourist development. These factors are considered below.

The following discussion is largely confined to tourist development in developing countries although many of the points have relevance to the developed world in a less extreme form. Some developing countries are already heavily involved in tourism and have considerable experience of its various manifestations. Many of these countries have exhibited marked economic successes. Turner (1976: 253) suggested that Mexico, for example, was able to avoid the industrial stagnation and inflation found in much of Latin America because of the buoyancy of her tourist industry. Tourist receipts permitted Mexico to import more than the other countries of that continent. By the 1970s tourism had also emerged as a major export industry in Spain, Greece, Kenya, Tunisia and Morocco.

In spite of persistent attempts to promote tourism, many developing countries have shown disappointing results. Diamond's (1977) case study of Turkey indicated that that country is typical of those developing nations which are endowed with tourist potential but whose resources are grossly underutilized and mismanaged. Some countries have only rudimentary tourism and others are still considering its possible development (Popovic 1973: 187).

If tourist attractions are appraised on a global scale, it is evident that developing countries are often richly endowed with outstanding tourist assets.

Popovic (1973: 189), writing on East African tourism, predicted that once the tourist attractions are better known, and sufficient facilities are created for less expensive travel and a more comfortable stay, they will enjoy an important share of the world's international tourism. The natural resources of these countries are very appealing to the North American and European visitor: wildlife, coastlines, mountain and lake scenery and, above all, their amenable climate. The non-reproducible resources of climate and ocean beaches are essential ingredients in the provision of 'sunlust' tourism (Gray 1974: 387) and have already contributed to the growth of Caribbean, Spanish and Greek tourist industries. Parsons (1973: 129) noted, in reference to Spain, that the advent of tourism based on low-cost charter air travel and rigorous promotional programmes, has enabled greater numbers of summer migrants to travel than were doing so previously. As a result the resorts of the Costa Brava and Costa del Sol have become as well known to Europeans as the French and Italian Riviera and the historic and cultural centres of the continent's capital cities. The same could happen in other developing countries, although not all are as conveniently located with respect to potential markets as Spain. The European and North American tourist searching for sun and the sand of warm beaches, but wishing to avoid crowding and congestion, may seek alternative destinations. Hence many developing countries stand to gain both from their possession of natural resources which are in great demand, and from the social and environmental problems resulting from the crowded conditions found in many highly developed resorts.

The labour requirements of the tourist industry are often especially suited to conditions prevailing in developing countries (Mings 1969: 176). One of the first tasks of economic development is to find gainful employment for all people. Developing nations are usually characterized by high unemployment and the ability of tourism to use labour intensively is an important virtue of the industry (Gray 1974: 395). From an input-output study undertaken in Mexico, Bond and Ladman (1972: 46) noted that forty-one jobs were created by an investment of US $80,000 in tourism. This was twenty-five more than would have been created by the same investment in the petroleum industry, and twenty-six more than in metal products.

When compared with many other industries, tourism requires employees with relatively low levels of job specialization. Thus, it may be possible to absorb a large proportion of the work-force from traditional sectors of the economy with a minimum of training. In the Mexican hotel industry, 50 per cent of jobs were non-specialized, and of the 42 per cent which were specialized (this excludes management and top-level administration), a large proportion only required a small investment in the training of personnel. However, Gray (1974: 395) argued that the ability of tourism to use large amounts of unskilled labour is only a temporary phase in the development of the industry. As the industry expands it may become more reliant on labour with higher skills. This could create labour shortages and lead to higher costs per unit of labour output. As other sectors of the economy expand, they will also increase the average level of skill required of the working population, and this will necessitate greater quantities of capital formation to sustain levels of economic development. Nevertheless, Gray argued that this should not be a deterrent to the promotion of tourism.

One of the appeals of tourism as a vehicle for economic development lies in the rapid rate of growth in numbers of tourists emanating from developed nations, and in the expectation that increased affluence in these nations will be reflected in faster rates of tourist generation (Gray 1970: 131). As total pleasure travel may expand with increased affluence, domestic capacity may become increasingly small relative to total demand and the ratio of foreign to domestic travel may greatly increase. Travel expenditures in developing nations are likely to benefit the most from this process.

A further, more mundane, factor is that much of the publicity given to tourism in relation to economic development by governments is a reflection of their strong commitment to the travel industry. Competition between government departments for financial allocations and subsidies favours those who can promote a growth industry. There is, of course, a danger of exaggerating the potential of tourism and this will be discussed later.

In summary, the availability of undeveloped resources, the nature of the labour market, and the growth of the international travel market may give developing countries some advantages over more developed economies for the development of tourism.

Advantages of tourism

What is there about tourism that gives it more appeal than other avenues of economic growth in developing countries? This question can be addressed through a consideration of the disadvantages of trading on the international market primarily through the export of primary produce (raw materials and foodstuffs), which is a characteristic feature of most developing economies. According to Bond and Ladman (1972), there are four major disadvantages:

1. The price obtained for raw materials is governed by the world market price and is subject to terms-of-trade conditions;
2. The export commodities of developing countries are usually concentrated upon a limited number of products;
3. Export markets in raw materials are unstable and, therefore, foreign earnings are uncertain;
4. Export of raw materials is conducive to a high propensity to import manufactured products in order to meet changing and increasing consumption patterns.

Tourism, as an invisible export, is not subject to these disadvantages to the same degree as many primary products.

Tourism also has a number of comparative advantages when compared with the export of primary products. The tourist exporting country (i.e. the host nation) has a greater degree of control in establishing prices for tourist goods and services than it does for the export of raw materials. In gaining such control the developing country is not subject to the vagaries of commodity exports which tend to be intensified in times of economic instability or decline (Ball 1971: 23). In this situation countries are able to manipulate prices within the tourist industry as a way of providing incentives for foreign travellers to visit. Developing countries involved in tourism are also favoured by the fact that tourism is highly income-elastic when compared with raw materials. Thus, slight increases in the incomes of potential visitors can lead to appreciable growth in tourist arrivals and receipts. On the other hand, larger increases in income are required to generate equivalent expansion in export earnings from the sale of raw materials. There is every indication that these relationships between income elasticity and sales will continue.

Secondly, tourism, by complementing other export products, adds diversity to the export base of a country and, thereby, helps to stabilize its foreign exchange earnings.

Thirdly, tourism has the potential to furnish foreign exchange to offset deficits created by growing demands to import scarce raw materials and manufactured goods. Tourism generally requires relatively little, by value, in imports for every unit of foreign exchange which it generates. Thus a greater proportion of the foreign exchange earnings of tourism can be used for investment in the development of manufacturing industries or in reducing the foreign earnings debt. The extent to which this is possible depends upon the ability of a country to supply the tourist industry from domestic rather than imported sources.

In theory, then, tourism offers developing countries considerable potential for economic growth. The degree to which tourism is an agent of development depends upon the characteristics of the country, the identification of realistic goals which are in line with these characteristics, and the ability to achieve these goals. Successful tourism development can only take place where the necessary prerequisites, or conditions, exist for the goals to be achieved.

Conditions for development

Wolfson (1967: 50), among others, has suggested that there has been a tendency for administrators in some developing countries to view tourism as an easy means of economic development. She concluded that this is far from the case. Tourism will only flourish given the appropriate conditions. It is an industry which, like any other industry, requires sophisticated planning and organization if its full potential is to be realized.

It has already been pointed out that destinations may be readily substituted by potential tourists as competition within the industry intensifies. Turner (1976: 254) noted that tourists who originally went to Spain began to seek alternative destinations in the 1970s and moved their allegiance to Greece, the Seychelles and North Africa. He also indicated that this meant a considerable reduction in per capita profits. For example, in Greece the average spending per head was US $126.50 in 1961 but by 1970 this had dropped to US $120.30 in spite of an increase in the average length of stay. Many of the primary resources of developing countries which initially attract tourists are somewhat similar. Therefore the ability of destinations to compete globally depends largely upon the four following conditions:

1. The mixture, quality and prices of the facilities and services being offered;
2. The existence of a skilled and experienced organizational body;
3. The geographical location of the destination area in relation to the main tourist generating regions, and the ability of these destinations to capitalize on the advantages of being well located, or to ameliorate the disadvantages of being poorly located;
4. The nature and origin of financial investment.

The first condition is that developing nations incorporate elements of stability within their development plans. Stability can be enhanced through the provision of a diversity of facilities and services which cater to a number of tourist types. These facilities must be comparable in quality and competitive in price with those of other destinations. Where a large proportion of tourist arrivals come from charters, tourist facilities and services must have the capacity to absorb large numbers of visitors. However, assuming that tourism is subject to large seasonal fluctuations in demand, it can be economically disastrous if excess capacity is created. The gains of additional patronage in the tourist season can be negated by gross underutilization of facilities in the off season. In the case of charters, it is imperative that supporting infrastructural and service requirements are met to ensure the successful continuation of the package. The availability of local excursions and tours, *en route* shops, restaurants and souvenir stands, and high quality water and sanitation facilities are just as important as luxury hotels constructed at the main entry and departure points. In other words, the total tourist package should meet the expectations and standards of tourists.

The likelihood of there being a co-ordinated set of complementary services and facilities will be enhanced by the existence of a skilled organizing, planning and marketing body. This is a second condition for successful tourist development. Mitchell (1970: 9) believed that one of the major constraints on recent expansion of tourism in East Africa has been the sluggishness of the government in simplifying the construction of attractive but affordable facilities. It is generally agreed that a government office of tourism is often the most appropriate organization: 'to regulate and control the quality of service. In marketing the product, effective economies of scale can be achieved through government advertising rather than individual effort. This marketing approach considerably reduces the number of skilled personnel needed to put the industry in contact with its customer' (Bond and Ladman 1972: 50). Government may also play a direct role in management and investment, particularly where tourism is a principal element of development plans (Wolfson 1967: 51). However, care should be taken concerning relationships between the public and private sector. Government activities should not preclude the involvement of the private sector, nor should the efforts of government be excluded, even where the private sector is active and experienced. Organizational bodies at all levels should attempt to co-ordinate developments so that opportunities for tourists are expanded and returns to the destination area are maximized.

The functions of a government organization, such as a national tourist board, may vary from public relations and promotion to market research, the preparation of development plans, direct financial investment and the operation of facilities. Popovic (1973: 194) suggested that the improvement of facilities for the training of local personnel for skilled jobs and for high-level management and executive positions should be one function of national tourist boards. He suggested that education for the latter role might be most beneficial if undertaken on a macro-regional level which would involve the co-operation of several national organizations. The importance of co-operation between national organizations was also stressed by Mitchell (1970) in his writings on East African tourism. He cited three major gains from this form of co-operation:

1. Activities such as tourist research, planning, promotion and education, and infrastructural facilities such as airports, roads and game parks all exhibit substantial economies of scale. It is suggested that gains from the provision of these services and facilities will be higher if they are organized and financed on a regional basis, rather than by individual governments;
2. The closest competition for tourist receipts among East African countries is from their East African neighbours. If individual countries of the region engage in outright competition with each other, this could lead to undercutting, and it is likely that gains would be less than if they co-operated. It is not suggested that competition be removed completely as it does stimulate the provision of quality facilities and services. Mitchell (1970: 14) argued that in the short term co-operation redistributes benefits but at the expense of jeopardizing receipts to any one country. In the long term co-operation will be to the benefit of all;
3. The removal of administrative obstacles to travel makes touring more convenient and less costly. This facilitates the creation of package tours with attractive schedules which are conducted on a regional basis and can encompass the tourist attractions of more than one country.

The geographical location of resources and markets has received a great deal of emphasis in research on the distribution of industry. The topic has received much less attention in the context of tourism. However, this does not reflect a lack of importance of location for tourism development and, in fact, a suitable location is a major prerequisite of a successful tourist industry. The cost of transportation to and from the destination is a major expense in tourist packages and, therefore, the location of the tourist exporter in relation to tourist-generating countries is a significant factor in the total vacation package cost structure. The demand for travel to one resort depends upon the cost of travel to and from that resort and the costs of travel to competing suppliers (Gray 1974: 388). Location close to large markets, and its bearing on transportation costs, was an obvious influence on the growth of Spain,

Mexico, Italy and Hawaii as major tourist destinations. Prior to the advent of long-haul jet aviation, it was those resorts closest to the tourist-generating countries which grew most rapidly (Turner 1976: 254). The high costs of multi-stop flights to more distant destinations was a major factor which hampered their development. Evidence recently presented by Sunday (1978: 268) indicated that the demand for travel in America is still sensitive to changes in air fares. He concluded that, other things being equal, more distant countries will receive fewer tourists than more proximate ones.

Many developing countries lie at a great distance from the principal tourist-generating countries of Europe and North America. Therefore it is vital that developing countries sell their tourist products to them at very competitive prices. These prices must be sufficiently attractive to facilitate competition with resorts within the tourist-generating continents, and with their developing nation counterparts. As many developing countries do not own their own airlines, they must rely upon foreign carriers to service their countries. Both the tourist receiving countries and the airlines want to derive full benefits from tourist travel, the former favouring low prices and the latter often having a preference for higher prices. Developing countries in this situation are at a comparative disadvantage. Unless an agreement is made between the receiving country and the air carrier, the former may have to alter the selling price of the tourist product, usually forgoing profit in doing so, to counteract the high costs of transportation and still be competitive. The degree of success of developing countries in offsetting the disadvantages of distance is strongly influenced by co-operation with the airlines. A number of possibilities exist for stimulating co-operation:

1. By receiving countries investing in the airline company or the airline company gaining a vested interest in tourist development within the destination; for example, by investing in hotels;
2. By receiving countries subsidizing the costs of travel through direct payments, or by taking the responsibility for financing marketing and promotional programmes. This is particularly appropriate in the case of charter air travel;
3. By liberalizing entrance formalities;
4. By individual countries with developing economies combining to form or expand their own airlines or to facilitate the other co-operation possibilities.

A fourth condition determining the success of tourism as a promoter of economic development is the nature and origin of investments. Most developing countries require more accommodation facilities, improvements and extensions to infrastructure, parks, and upgrading of the quality of tourist services. These can only be achieved with substantial financial investments. Mitchell (1970: 9) established a high capital–output ratio of between 2.5 and 3.0 for tourism in Kenya. (A capital–output ratio of 3.0 implies that for every three units of capital input, one unit of output emerges.) Fifty per cent of the capital was required for investment in hotels, tour operations, and local air charter firms; 30 per cent for infrastructural developments such as roads, airports, public utilities and game reserves; and the remainder for training programmes and the support of miscellaneous industries contributing to tourist attractions.

The nature of financial investments is as important as the amount. There is no doubt that favourable investment opportunities exist in tourism for both the public and the private sectors. Long-term investments can be particularly rewarding in functions directly serving the tourist market, such as accommodation. Nevertheless, private investors have been reluctant to invest in the creation of tourist accommodation in developing countries. This has been because of the seasonality in demand and because of the reluctance of potential investors to tie up large amounts of capital in fixed assets for long periods of time. Governments of both developed and developing countries, and institutions such as the World Bank, have attempted to overcome this problem by the provision of subsidies, tax and duty concessions, and by the availability of attractive long-term loans. The World Bank, in particular, has recognized the potential of tourism in developing countries by investing in East Africa, Yugoslavia, Tunisia, Morocco, Nepal and Mexico. Investment in the form of long-term loans is highly desirable to destination areas because it provides an opportunity for local entrepreneurial activities and keeps most of the profits within the tourist exporting country.

Investment in developing countries may also be in a more direct form. International hotel chains, car hire firms, tour operators and food chains frequently establish their own operations. It is not uncommon to find Hertz and Avis rent-a-cars, Holiday Inn and Sheraton Hotels, and McDonald's and Coca-Cola in well established resorts of developing countries. Although foreign investments of this type remove the demand for capital from the host country and permit the redirection of local capital to other avenues of investment, it is questionable whether they contribute greatly to local profits. Most of the benefits accrue to shareholders in the developed world. It is important that provision be made within the economic structures of tourist-exporting countries for indigenous investment and employment, and for purchasing policies which encourage a gradual replacement of foreign control of the economy.

The above discussion has examined some of the ingredients which determine the success of tourism as an economic development strategy in developing countries. In theory, tourism can make an important contribution to the early phases of economic development. Its role is likely to diminish in significance as industrialization takes place, and as labour and production costs increase. Tourism should not be viewed as the principal, long-term source of foreign exchange, and profits from tourism should be directed into other sectors of the economy.

The realities of tourism and economic development

Measures have yet to be developed which indicate the performance of tourism as a contributor to economic development. Frequent mention is made of the role of tourism in generating employment and in assisting in the reduction of balance of payments problems. However, these are not measures of overall economic growth, although they may be factors which are significant to it. In the absence of a universally acceptable methodology for evaluating the performance of tourism as a catalyst of economic growth, it is difficult to draw conclusions on the topic. Mixed opinions have been expressed by other authors. Failures, or only marginal successes, have been noted for the West Indies by Perez (1974), for the Languedoc-Roussillon region of Mediterranean France by Willis (1977), and for Turkey by Diamond (1977). In contrast, success has been documented for Mexico by Ball (1971), Bond and Ladman (1972) and Jud (1974), and for India by Krishnaswamy (1978). These examples reveal a diversity of experiences with tourism as an instigator of economic growth. They also point out a variety of demand and supply problems which will be illustrated by reference to a number of case studies.

A case study comparison

Mexico has had considerable success in its use of tourism as a stimulus to economic growth. Jud (1974: 30) postulated that tourism has promoted economic growth in Mexico in two ways:

1. It has created employment and absorbed part of the labour force left redundant by the mechanization of agriculture;
2. It has eased foreign exchange constraints upon development and has reduced fluctuations in foreign exchange earnings.

On the other hand, tourism in Turkey (Diamond 1977) and in the West Indies (Perez 1974) has contributed little to the economic development of these regions. In Turkey, for example, tourism not only failed to create much employment, it also required types of labour which were in short supply. The fragmented nature of the industry meant that large numbers of people were needed to supervise but management skills were in short supply. In spite of high rates of investment in the industry, tourism in Turkey only generated 4.3 per cent of total exports, by value, which was low when compared with nearby Greece (13 per cent) and Yugoslavia (10 per cent).

A number of factors contributed to these differences in performance. Factors related to tourist demand included the following:

1. *The level of competition.* Mexico began its tourism boom in the 1950s and was well established in tourism by the time Turkey, and many other developing countries, began to venture into tourism. Those countries which were slow in entering the market had to face more intense international competition;
2. *The location of the host country in relation to its main source of tourist arrivals.* Mexico is well placed to receive tourists from North America, while Turkey is poorly placed in this respect. Turkey's major asset is its amenable climate and its main generating countries are in north-west Europe. Unfortunately for Turkey, there are numerous 'sunlust' facilities between it

and north-west Europe and, as a result, Turkey has had to wait for overloading at the intermediate resorts before receiving an appreciable increase in tourist arrivals;

3. *The instability of demand.* Both Mexico and Turkey experience seasonal fluctuations in arrivals, but Mexico is less subject to problems of seasonality than Turkey. This is because of the close proximity of Mexico to North America and its special attractions of historical sites, cheap shopping and gambling. Turkey has been subject to internal political uncertainty and a succession of natural disasters, such as earthquakes and floods, which act as deterrents to potential visitors.

Collectively, these factors have contributed to the development of different demand structures which have significantly affected the growth of each nation's tourist industry.

From a supply perspective, there is no evidence to suggest that Mexico has experienced particular difficulties. Turkey, on the other hand, has experienced considerable supply problems and is an instructive example from which other countries could learn. These problems include the following:

1. A high capital intensity which has been caused by heavy infrastructural requirements and large fixed capital investments in accommodation;
2. High import requirements, both in construction and recurrent inputs, increased costs of development;
3. A small proportion of inputs were obtained from other sectors of the domestic economy;
4. Poor transportation and communication facilities;
5. A low level of organization;
6. A high proportion of profits were returned to the tourist-generating countries.

The Turkish example is an explicit illustration of the problems faced by developing countries as they attempt to expand their tourist industries. Perez (1974: 475–6), writing in a different locational context, summarized these problems and pitfalls as follows:

> The West Indies lacks the resources to support the volume of tourists and the concomitant vacation life-style metropolitan agencies impose on the region .... The large import component necessary to support tourism ... serves at once to sustain metropolitan growth and foreign imports while reinforcing underdevelopment; imported materials, food prepared abroad and expatriate staffs make up the 'invisible' support system accompanying the traveller to the region ... expansion of tourism ... exacerbates the chronic balance of payments deficit and tourist generated employment is reminiscent of the colonial monoculture systems.

This discussion has addressed the question of the contribution of tourism to the economies of developing countries. Ghali (1976: 538) asked a further question: was it worth it? The following chapters will provide some insights into this question. Some authors have regarded tourism as a disguised form of economic imperialism. The weakness of this perspective is that it stigmatizes the whole industry which, given the right circumstances, has the potential to yield enormous benefits. Tourism does have its economic shortcomings, but it also has merits which can be harnessed to the benefit of destination area economies. The next section of this chapter examines the components of economic benefits and economic costs. It is the sum of these components which collectively determine the contribution of tourism to economic development.

Economic benefits and costs

The magnitude of economic impact is governed by a multitude of factors. Some of the more pertinent ones include:

1. The nature of the main facility and its attractiveness to tourists;
2. The volume and intensity of tourist expenditures in the destination;
3. The level of economic development of the destination area;
4. The size of the economic base of the destination area;
5. The degree to which tourist expenditures recirculate within the destination;
6. The degree to which the destination has adjusted to the seasonality of tourist demand.

The above factors also determine whether the economic impact is positive or negative.

The economic benefits of tourism, which have been documented in the literature, include the following:

1. The contribution of tourism to foreign exchange earnings and the balance of payments;
2. The generation of income;
3. The generation of employment;
4. The improvement of economic structures;
5. The encouragement of entrepreneurial activity.

Much less is known about the economic costs of tourism than the benefits. Costs mentioned in the literature include:

1. The danger of overdependence on tourism;
2. Increased inflation and higher land values;
3. An increased propensity to import;
4. The seasonality of production and the low rate of return on investments;
5. The creation of other external costs.

Tourism and the balance of payments

The potential contribution of tourism to the balance of payments as an earner of hard currency has been widely recognized. Few countries have escaped balance of payments problems over the last decade so that any industry that is likely to generate foreign earnings has been liable to receive the support of government through the provision of incentives. This, according to Young (1975: 43), has often occurred at the expense of other industries which have a high import content or are not export oriented. Many countries with a strong dependence upon international trade have been heavily influenced by balance of payments considerations in formulating official policies regarding tourism. For example, the United Kingdom Tourism Act of 1969, which is no longer in force, placed an upper limit of £50 on spending money taken out of the country.

The balance of payments has emerged as one of the most publicized of all the economic considerations surrounding tourism. Most of the reports published by the International Union of Official Travel Organizations, the US Travel Data Center, and governments or national tourist organizations have been primarily statistical in orientation. There have been few research papers which have examined theoretical relationships between tourism and the balance of payments to identify the ways in which the effects of tourism are felt and recorded in the balance of payments account. The early studies by Lickorish and Kershaw (1958), Peters (1969), and Gray (1970) gave accounts of the volume of international tourist receipts and their increasing proportion of the total value of world exports, but only brief mention was made of the balance of payments effects of international tourist expenditures. However, more recent studies have been more critical in their examination of balance of payments questions and have been more penetrating in their analyses and evaluation of available data (Erbes 1973; Thuens 1976; Airey 1978).

Table 2 Structure of the balance of payments

1 Current account	all current transactions including earnings and expenditure in goods and services
A. Merchandise	(trade) – visibles
B. Invisibles	*Transportation* includes all foreign earnings from home country owned carriers and their disbursements overseas, and home country residents expenditure on foreign carriers and their disbursements to the home country. *Travel Expenditure* personal expenditure in the home country and abroad. *Income on investments* interests, profits, dividends. *Private gifts* transfers in the form of gifts or family remittances. *Other services* financial transactions in banking, insurance and brokerage, advertising agency expenses, telecommunications.
C. Governmental	All transactions between the home country and overseas residents.
2 Capital movements	long-term and short-term capital transactions, investment flows, trade credits and other capital flows.
3 Gold movements	in and out

To arrive at a balance of payment a system of credit and debit items is established for each import-export type listed above.
Source: Samuelson 1967.

The balance of payments account for a country is a record of economic transactions during a period of time (usually a year) between residents of that country and the rest of the world. It takes into account the value of all goods, gifts, loans, foreign aid and gold coming into and leaving the country, and the interconnections between these items. The structure of the balance of payments has three sections, as indicated in Table 2. Tourist expenditures, both within the home country and overseas, form part of the current account. The contribution of any economic activity to the balance of payments consists of currency outflow sold to overseas residents, and the secondary and tertiary effects of that activity (Airey 1978: 8).

The effects of tourism on the balance of payments consist of two components:

1. The effects of tourism within the home country, including the country's own residents and visitors from overseas;
2. The effects of international tourism, i.e. the tourist activities of residents which take place outside of the home country.

Some authors question the necessity of separating the contribution of tourism in this way. However, it is important to know the effects of the tourist activities of foreign nationals visiting the country so that one can determine the role of foreign earnings in the balance of payments account. In some countries foreign earnings from tourism may be very low and domestic tourism may even be a drain on the balance of payments. The costs of servicing both foreign and domestic tourists may be so great, and the imported component so large, that earnings from foreigners may be insufficient to offset these high costs. Countries are also interested in the effects of tourism beyond their boundaries. They wish to know if more money is leaving the country than is being brought in by international tourism.

Airey (1978: 4–5) divided the effects of tourism on the balance of payments into three categories: primary, secondary and tertiary effects. Primary effects are direct, immediate and relatively easy to measure. They refer to actual visitor expenditures made by foreign visitors within the home country, and residents of the country abroad. They give rise to direct inflows and outflows of currency, respectively. Primary effects of tourism will only occur if travellers have crossed an international boundary. Primary effects, whether expenditures on accommodation, entertainment, shopping, transport or cars for export, are recorded and reported separately. This makes the assessment of primary impacts a relatively straightforward task.

Secondary and tertiary effects are more complex, more difficult to measure and, in consequence, have been left out of most balance of payments assessments. Given the importance of tourism in balance of payments issues, it is surprising that little attention has been given to these effects. Secondary effects are the effects on the balance of payments of the direct tourist expenditures as they percolate through the economy. Secondary effects, therefore, do not require the initial visitor expenditure to have taken place in another country; they may appear in a number of different forms:

1. *Direct secondary effects*. Expenditures on overseas marketing, visible imports, commissions to travel agents, outflows in the form of interest and dividend payments to overseas investors, and airline operator disbursements on their crew's overseas expenses are direct secondary effects. Airlines, hotel operators, and other tourist services have to import supplies from abroad to cater to foreign tourist demands and, therefore, incur payments to overseas suppliers. These are outflowing direct secondary effects;
2. *Indirect secondary effects*. Earnings gained by the initial tourist services will be passed on to other suppliers of goods and services. Their production, in turn, may depend upon a supply of imported goods whose purchase requires an outflow of currency from the home country. For example, an airline gives a domestic company the contract to supply the goods for on-board services. In doing so, it passes part of the initial tourist expenditure on tickets to a supplier of food and drink. They, in turn, may have to import certain foods or necessary goods and this creates an outflow of currency from the home country. This secondary transfer of initial tourist expenditures is known as an indirect secondary effect;
3. *Induced secondary effects*. As expenditures permeate through the economy they will generate payments to producers of tourist goods and services, and their employees, in the form of wages, salaries and rents. The proportion of the labour force which is of foreign origin will determine, to a large extent, the proportion of these payments which is remitted abroad.

The currency flows generated by tourism do not all constitute primary or secondary effects. Flows of currency not initiated by direct tourist expenditures are termed tertiary effects. These include: imported goods, for example clothes and suitcases, purchased by residents of the home country in preparation for travel, and investment opportunities created by tourist activity. The existence of tourist activity may stimulate home country companies to export specific commodities to tourist-generating countries and to import others. A recent example of this has occurred in New Zealand. The large numbers of Japanese and American tourists purchasing huge quantities of sheepskin products has not only directly stimulated the local industry, it has also prompted existing companies to export these products for sale in the retail markets abroad.

The effects at secondary and tertiary levels are not recorded separately and they are difficult to identify. Moreover, identification and measurement problems increase as money slowly filters through the economy. Description and classification of effects, as has been done here, is simple when compared with the tasks of measurement and assessment.

There is no doubt that many countries are suffering from large balance of payments deficits and are seeking measures to correct the situation. The United Kingdom, the United States and Canada are examples of such countries. In spite of the many criticisms of the tourist industry and, particularly, its activities in developing countries, most countries view the possibility of expanding the tourist industry with favour. Relationships between tourism and the balance of payments attract politicians and planners, and cause them to seek guidance on the viability of expanding the industry. The way in which the effects of tourism are presented in the balance of payments accounts has, therefore, considerable bearing on the justification of such decisions.

As mentioned earlier, a full account of the impact of tourism on the balance of payments should include secondary and tertiary, as well as primary, effects. However, most countries are not in a position to assess such effects because they are not recorded separately in the accounts. Although methodologies, such as economic multipliers (see later), have been developed to measure secondary effects, there have been few attempts to relate them to the balance of payments. As a result only primary effects, or the direct effects of overseas visitor expenditures, are usually examined. However, as Thuens (1976: 2) commented, these only represent gross earnings which should be weighed against the foreign exchange expenditures which are used to promote earnings from the tourist industry. He stated that: 'not the gross effect but the net effect on the balance of payments is the entity which with regard to the export of tourist services finally matters'.

It is natural to emphasize international tourist activity in assessments of the effects of tourism on the balance of payments. This includes expenditures made by overseas visitors in the home country, and by residents of the home country abroad. On many occasions these effects are considered together as the travel account, and are compared to see which is larger. In other studies a wider, though not comprehensive, approach is employed in which all readily identifiable items of international visitor expenditure are incorporated. These might include foreign investment patterns, and money spent on transport and the training of foreign staff. The result is known as the tourism balance. The travel account and the tourism balance are unlikely to be identical but the outcome is the same: neither indicates the true contribution of tourism to the balance of payments.

In spite of these reservations, the travel account may provide a useful preliminary assessment of the involvement of a country in international tourism.

Table 3 A hypothetical example of the relationship of the travel account and tourism balance to the balance of payments

Country	Balance of payments on current account* ($m.)	Travel balance† or account	Tourism balance‡
A	−250	−30	−5
B	+10	+40	+120
C	−400	+160	+250

* The current account is a recording of all current transactions and includes earnings and expenditure resulting from transactions with overseas residents in both goods and services.

† This is the balance between the expenditures of residents of country A overseas and of overseas residents in country A (i.e. the travel account).

‡ This is the balance between all identifiable items of expenditure by all overseas visitors to country A and all identifiable items of expenditure by residents of country A on travelling overseas.

This is illustrated by the hypothetical example which is outlined in Table 3. The travel account can accentuate balance of payments deficits (Country A) or surpluses (Country B), or may be a significant factor in reducing deficits caused by other transactions (Country C). Residents of Country A spend more overseas than foreign residents spent in their country and, thus, the travel account is negative. The situation was the reverse for countries B and C, whose travel accounts bear a positive sign. The incorporation of a second measure, the tourism balance, yields a more clear, though still incomplete, picture of the overall positive or negative economic effect. In the case of Country A, expenditures on such items as transportation have reduced the contribution of tourism to the deficit. The positive contribution is even more marked in B and C. The increases over the travel account of in excess of $80 million are likely to be the results of the earnings of air carriers and high levels of overseas spending within these countries. Tourism, in the case of countries B and C, has counteracted a small or negative balance of payments, but in country A it has been destabilizing influence.

The travel account is an indicator of the degree to which a country attracts overseas visitors when compared with its ability to persuade home residents not to travel abroad. The travel account makes no reference to secondary or tertiary effects occurring as a result of domestic tourist activity. The travel account, therefore, only provides part of the total picture. The balance on the travel account is calculated by subtracting the expenditures of residents travelling abroad from the expenditures by overseas residents in the home country. These are really separate activities and it may serve no real purpose to balance them. Gray (1970: 89) summarized the role of this balancing procedure when he wrote as follows: 'The practice of netting out exports and imports on an individual account is nothing more than an accounting convenience and the concept of the gap as a measure of the responsibility of an activity for the overall deficit is fallacious.' The ease with which the calculation can be made has contributed to the over-use of the balance on the travel account in assessing the contribution of tourism to the balance of payments. Perhaps more meaningful results would be obtained if these two effects were treated separately. If this were done, the effects of tourist expenditures could be compared with other forms of expenditure. For example, expenditures on tourism overseas could be expressed as a percentage of total consumer expenditures, or compared with other forms of overseas expenditures. These measures would provide an indication of the relative importance of tourism when compared with other activities in the economy (Airey 1978: 7).

In summary, in assessments of the effects of tourism on the balance of payments, most attention has been devoted to the primary effects of tourist expenditures. As a result, the scope of the data and the associated studies has been severely limited. The amount of money that is spent on tourism is one thing; where this money goes and the effects of its circulation are quite another. Attention, therefore, should be focussed on the wider implications of tourist activity, including the magnitude of foreign resources used in meeting domestic demands, as well as the expenditures of residents overseas.

Some examples

Given the rather abstract discussion of the balance of payments which has been presented, it is now appropriate to discuss some specific examples. However, of necessity, comments are made only on those effects for which information is readily available. Deficiencies in data and underlying methodologies should also be borne in mind.

Table 4 Trade balance and tourism balance, 1966–75 (in millions of US dollars)

	Year	External trade		Trade balance	International tourist receipts and expenditures		Tourism balance
		Exports	Imports		Receipts	Expenditures	
Italy	1966	8,038	8,589	−551	1,460.3	260.9	1,199.4
	1967	8,705	9,827	−1,122	1,423.7	297.6	1,126.1
	1968	10,186	10,286	−100	1,475.6	363.3	1,112.3
	1969	11,729	12,467	−738	1,632.4	492.8	1,139.6
	1970	13,206	14,970	−1,710	1,638.4	725.9	912.5
	1971	15,116	15,982	−866	1,882.3	836.9	1,045.4
	1972	18,548	19,282	−734	2,174.0	1,049.0	1,125.0
	1973	22,264	27,797	−5,553	2,372.9	1,458.6	914.3
	1974	30,253	40,924	−10,671	1,914.0	1,227.0	687.0
	1975	34,830	38,366	−3,536	2,578.8	1,050.1	1,528.7
Spain	1966	1,254	3,572	−2,318	1,292.5	90.5	1,202.0
	1967	1,307	3,453	−2,146	1,209.8	99.4	1,110.4
	1968	1,589	3,502	−1,913	1,212.7	101.5	1,111.2
	1969	1,900	4,202	−2,302	1,310.7	86.0	1,224.7
	1970	2,387	4,715	−2,328	1,680.8	113.0	1,567.8
	1971	2,938	4,936	−1,998	2,054.5	136.2	1,918.3
	1972	3,803	6,755	−2,952	2,607.6	190.4	2,417.2
	1973	5,161	9,537	−4,376	3,091.2	270.5	2,820.7
	1974	7,059	15,293	−8,234	3,188.0	325.7	2,862.3
	1975	7,691	16,097	−8,406	3,404.2	385.4	3,018.8

Sources: World Tourism Organization (1978), *Economic Review of World Tourism*. Madrid, p. 38
United Nations *Statistical Yearbook*. Various dates.

Table 4 presents balance of payments statistics for Italy and Spain. Both countries exhibit substantial surpluses on their travel accounts. In Italy tourism does not offset the deficit on the trade account. In contrast, in Spain in 1973, 1974 and 1975, the tourism balance reduced the trade balance by 64.4 per cent, 35 per cent and 36 per cent respectively.

Canada has had a moderately high deficit on the travel account in recent years (Fig. 3). The net balance of payments deficit on the travel account soared to $1,641 million (Canadian) in 1977, a 38 per cent increase over the previous year (*Statistics Canada* 1979: 154).

International tourism has become an increasingly important component of Britain's overseas earnings. Although expenditures overseas by British resi-

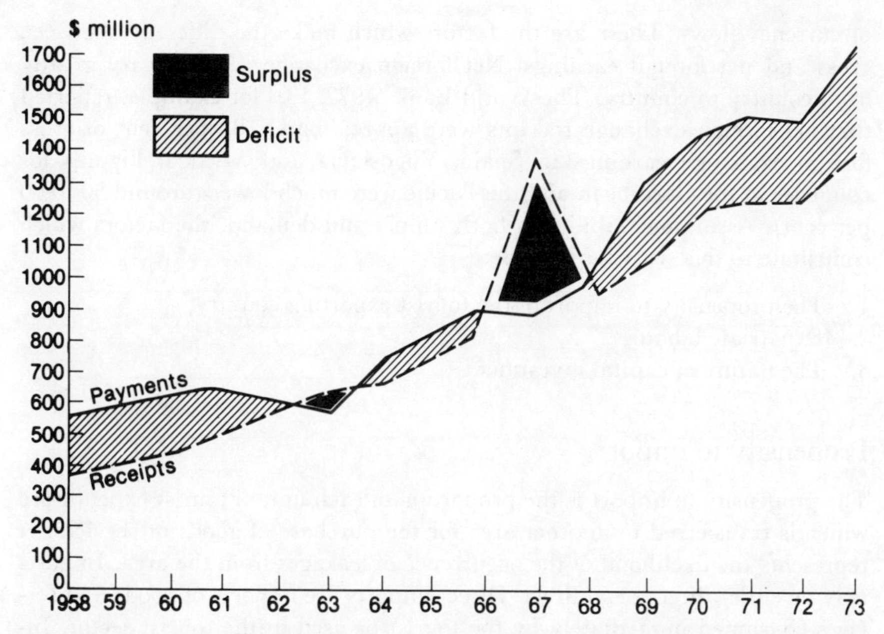

Fig. 3 Canada's travel account with all countries, 1958–1973 (*Source*: 'Canadian Tourism Facts Book 1972', Industry Development Branch, Canadian Government Office of Tourism, Ottawa, and Statistics Canada, Cat. No. 67–001. Found in Canadian Government Office of Tourism, 1974)

dents have increased in absolute terms over the past ten years, their significance within the accounts has fluctuated considerably. In spite of this, the positive tourism balance of the last five years has acted to reduce, but not eliminate, the balance of payments deficit.

Curtailment of foreign travel may be an attractive option for countries with a balance of trade deficit, particularly if the travel deficit is a significant proportion of the total. Some countries have imposed restrictions on foreign travel, and others have attempted to increase foreign tourist arrivals as a means of overcoming trade deficit problems. Gray (1970: 107–29) has considered the arguments for tourist import curtailment and export expansion in some detail. However, manipulation of the travel balance or gap is of only marginal value for countries with a large overall balance of payments deficit problem. Gray (1970: 88–9) summarized the reasons for this when he stated:

> The fact that the overall deficit must be reduced by some real adjustment . . . must be brought home to the general public . . . and travel imports are a possible source of foreign exchange savings . . . . Any measures taken to reduce the travel gap will, if they are successful, also tend to reduce any deficit in the balance-of-payments (ceteris paribus) but the overall effectiveness of these measures will be grossly exaggerated by the reduction of the deficit on foreign travel account.

Currency flows

Given the nature of the effects of tourism on the balance of payments, it is important to note the factors which determine the direction and magnitude of currency flows. These are the factors which make the difference between gross and net foreign earnings. Net foreign exchange receipts vary greatly from country to country. The World Bank (1972: 13), for example, reported that net foreign exchange receipts were approximately 85 per cent of gross foreign exchange earnings in Spain, Yugoslavia and Mexico. Figures for countries in the Caribbean and the Pacific were much lower (around 50 to 60 per cent). Assuming stability in both supply and demand, the factors which contribute to this situation include:

1. The propensity to import of the tourist exporting country;
2. Expatriate labour;
3. The nature of capital investment.

Propensity to import

The propensity to import is the proportion of each unit of tourist expenditure which is transferred to another area for the purchase of goods or services. It represents the likelihood of the occurrence of leakages from the area. Imports may be either direct or indirect. Direct imports are imports of goods and services consumed immediately by the tourist or used in the tourist sector. Indirect imports are the imports of raw materials, manufactured goods and services for domestic producers who provide goods and services to the tourist sector. The volume of imports will depend upon the extent to which the demands for these goods and services can be met domestically (Thuens 1976: 4). In many cases, particularly in developing countries, the local economy lacks the capacity and diversity to meet the requirements of international tourism. Bryden (1973: 33) reported that declines in agricultural production in the islands of the Caribbean during the 1960s, and expanded tourist demands, reduced self-sufficiency in food production and expanded the need to import food by up to 4 per cent annually. This meant an increase in the proportion of tourist expenditures used to purchase imported goods and services.

In addition to imports of food, the tourist industry frequently purchases specialized management and operating supplies, including beverages, equipment and construction materials, from abroad. In developing countries,

Table 5 Import propensities by type of establishment

Type of business establishment	Import propensity (%)
Food stores	49*
Liquor stores	66
Clothing and accessory stores	44
Hotels	38
Restaurants	41

* 49 cents in every dollar of tourist expenditure on food was used for the importing of foodstuffs.
Source: After Lundberg 1972

where there is a limited output of both agricultural and manufactured products, a large proportion of these requirements must be imported. Lundberg (1972: 137), using Hawaii as an example, indicated that different types of establishments have different import propensities. The overall import propensity for Hawaii is 45 per cent, and for Kenya 22 per cent, but Table 5 shows that there is considerable variation from establishment to establishment. The measures of import propensity by establishment type are more useful than a single figure calculated on a national basis. They can be used to indicate variations in the prospects for import substitution and, hence, as an aid in determining priorities in the development of domestic production.

The propensity to import is influenced by the size and development status of the importing country, and by its import policies. Some countries attempt to restrict imports by the erection of tariffs. However, no examples have been located of policies specifically devised to curb imports for the tourist sector. The economies of developing countries are frequently unable to supply the quantity and quality of goods and services required to meet the demands created by international tourism (Sadler and Archer 1974: 5). Developed countries, because they usually are more able to construct, equip, supply and operate their tourist industries from domestic resources, tend to have low import propensities when compared with developing countries. In developed countries the tourist industry is supported by a sophisticated system of backward linkages, which is only possible in a mature, diversified economy. The size of the country is also important. In small countries the economy is likely to be less diversified than in larger countries. In small, developing countries, there is usually a small ratio between net and gross receipts from international tourism, because of the need to import a large proportion of tourist-related products.

Expatriate labour

Employment of expatriate labour in the tourist industry is usually a result of the inability of the home country to supply the labour domestically. Tourism is said to require a higher proportion of semi-skilled and unskilled labour than many other industries, so that one might expect that developing countries entering the industry would not require the services of much expatriate labour. Thus they should be able to avoid the outflow of currency through remittance payments. This may be so in some cases but, in the Caribbean, quite the contrary has occurred. Bryden (1973) undertook a detailed analysis of tourism in the Commonwealth Caribbean and concluded that the proportions of expatriates employed in the British Virgin Islands, the Bahamas and the Cayman Islands were particularly high. Concentrations were greatest in managerial and administrative occupations. For example, in 1970 in the Cayman Islands, nearly 65 per cent of employees in these occupations were expatriates. In the British Virgin Islands, 48.5 per cent of the labour-force in hotels and guest-houses was from abroad. Although this figure was only 32.1 per cent for the Caymans, 42.7 per cent of the total wage and salary bill of the hotel and guest-house sector accrued to expatriate labour (Bryden

1973: 130). The proportion of the expatriate earnings which is remitted out of the country is unknown. However, the higher the total expatriate earnings, the larger the volume of leakage from the tourist exporting country is likely to be.

There is little documentation of measures to reduce such outflows of money. Immigration restrictions are one obvious remedy. Increases in the allocation of financial aid for the training of locals should also reduce foreign employment and remittances in the long term.

Capital investment

Foreign dominance in investment in and ownership of tourist plant is a common feature of tourism, particularly in developing countries (Jackson 1973; Marsh 1975b; Young 1973; Talbot 1974; Turner 1976). According to Turner (1976: 255), foreign control of the tourist industry is the normal situation, but it is also undesirable, especially from the perspective of developing countries (Perez 1974: 480). Foreign control of the tourist industry results from two major conditions:

1. *The encouragement of foreign investment in the early stages of tourist development.* The high capital requirements of infrastructural and service facilities force many developing countries to seek financial assistance abroad. Remittances, in the form of capital repayments and interest on investments, can be very large;
2. *The emergence of multinational corporations.* The foreign ownership problem has been compounded by the emergence of international hotel chains, tour operators, restaurant chains, and the tendency towards both horizontal and vertical integration in the industry. These trends have increased the power of the airlines and tour operators. For example, airlines have developed hotel chains, and tour operators have invested in hotel chains and even developed their own airlines. Young (1973: 71–110) cited numerous examples of this phenomenon. Clarkson's, one of the world's largest tour operators, owned the Court airline; Canadian Pacific owns airlines, railways, hotels and tour operators; Pan American airlines owned Inter-Continental Hotels. Vertical integration may increase efficiency but it often also means increased foreign control. It may result in consistently high occupancy rates in the hotels of developing countries but, since a large proportion of the revenue and profits is transferred back to the tourist-generating countries, it only brings marginal returns to the host country.

High leakages have contributed to the disappointing economic performances of the tourist industry and for its failure to eliminate large balance of payments deficits. High leakages, of course, are not restricted to the tourist industry but are common to many forms of modernization in developing countries. However, the existence of high leakages may mean that tourism, in its present form, is doing less than it might to reduce balance of payments problems in the developing world.

Tourism and the gross national product

The significance of tourist expenditures to the economic performance of a country can also be assessed through the gross national product (GNP). Mitchell (1970: 2) suggested that GNP is one of the more important measures of the pecuniary consequences of tourism.

The contribution of tourism to GNP is calculated by assessing visitor expenditures at current prices, for both domestic and foreign tourism, and then subtracting the goods and services purchased by the tourist sector (Lawson and Baud-Bovy 1977: 21). If desired, the calculations can be undertaken for foreign tourism alone. Payments made in a country by foreign tourists are recorded as foreign exchange receipts. The amount of GNP generated by foreign tourist expenditures consists of this income less the costs of servicing these tourists. For those countries which do not have a large domestic tourist industry, this measure may be of more interest than a figure for domestic and international tourism combined. This is particularly true of developing countries where domestic tourism may be small and international tourism can make substantial contributions to the economy. Unless tourism has become a major earner of foreign exchange, the output of the tourist sector is likely to constitute only a small proportion of the GNP. This applies to almost all developed and many developing countries.

Table 6 Percentage share of international tourist receipts in the gross national product 1973–1975

Country	1973 (%)	1974 (%)	1975 (%)
Australia	0.3	0.3	0.3
Austria	8.0	6.9	7.4
Canada	1.2	1.0	1.0
Germany (Fed. Republic of)	0.6	0.6	0.7
Ireland	3.2	3.0	2.8
Japan	5.1	5.2	5.1
Spain	4.4	3.7	3.4
Switzerland	3.4	3.0	3.0
United Kingdom	1.0	1.1	1.0
USA	0.2	0.3	0.3

Source: World Tourism Organization, *Economic Review of World Tourism* 1978.

In most developed countries it is necessary to include domestic tourist receipts because they form a large proportion of total tourist revenue. Peters (1969: 29) noted that international tourist receipts in the United Kingdom constituted a mere 0.6 per cent of GNP in 1966. However, when the expenditures of Britons within the United Kingdom and foreign payments to British carriers were added, the figure increased to 2.4 per cent. Similarly, international tourist receipts constituted approximately 1 per cent of GNP in Canada in 1976, but this rose to 5 per cent when the contribution of domestic tourism was included. In Table 6 the contributions of international tourist receipts to the gross national products of ten countries are compared. In most developed countries tourism makes only a modest contribution to the GNP but it rises somewhat in such countries as Austria and Japan where tourism is proportionately more important.

The net earnings of tourism do not accurately indicate the net impact of tourism, even in a narrow, economic sense. Mitchell (1970: 2) noted that local resources and capital are used to satisfy the needs of tourists. These resources and capital could be used in other sectors of the economy if they were not employed in tourism. These costs would need to be deducted to arrive at the net impact of tourism on GNP. There are obvious difficulties in meeting this requirement. Nevertheless, the contribution of tourism to GNP is a useful measure of the economic significance of tourism. It is also possible to compare tourist receipts to national income or total exports. Each measure indicates the importance of tourism in a slightly different way.

Economic multipliers

The emphasis of the preceding discussion has been on the economic impacts of tourism at a national level. Many of the data have been collected at this scale and a large proportion of the literature concerns individual countries. An equally important concern is the contribution made by tourism to specific destinations at the local and regional scales. Expenditures of tourists in a destination create new incomes and outputs in the region which, in turn, produce further expenditures and incomes. This process can be illustrated through a simplified, hypothetical example. Tourists spend money to fill up their cars with petrol. The petrol station operator must pay for his supplies of petrol which are brought in from outside the region so that some of the money leaves the region as leakage. If operating expenses are ignored, the remainder of the initial tourist expenditure is income to the petrol station owner. He, in turn, buys food from the local store. Again some money leaks out to pay for imported produce but some remains as income to the store owner. As long as a demand for locally produced goods and services exists, each successive round of spending will generate new income. The process of respending continues until all the money has leaked away. The respending of incomes, thereby creating additional incomes, is known as the multiplier effect. The tourist multiplier can be defined as the number by which initial tourist expenditure must be multiplied in order to obtain the total cumulative income effect for a specified time period.

The size of the tourist multiplier will vary from country to country, and from region to region, depending upon the nature of the economic base. The volume of imported goods and services consumed by tourists, the inclination of residents to use goods and services from outside the region, and their propensity to save, each have a bearing on the multiplier coefficient. The higher the region's propensity to import, the higher will be the leakage of money out of the local region, and the lower will be the multiplier. In general, the smaller the economic base, the less self-sufficient will be the region, and much of the tourist expenditure will be respent outside the local region leading to a low economic multiplier. The multiplier effect is also influenced by the internal structure of the economy and the manner in which the injection of tourist expenditures is distributed across its various sectors. The greater the internal linkages between sectors, the less likelihood that supplies will be required from outside the region, and the larger will be the multiplier.

Archer (1976: 115) defined the income multiplier as the ratio of direct, indirect and induced changes in an economy to the direct initial change itself. Thus the income multiplier considers three types of influence of tourist expenditures:

1. *Direct spending*. The initial expenditure creates direct revenue to hoteliers, service stations, and other tourist industries;
2. *Indirect spending*. The payments of salaries and wages to local employees, and tourist establishments replenishing their stocks, are indirect effects of the initial, direct tourist expenditure;
3. *Induced spending*. As wages and salaries within an economy rise, consumption also increases and this provides an additional impetus for economic activity.

The concept of the income multiplier is often confused with other types of tourism multiplier, with a resulting misinterpretation of their implications. In addition to the income multiplier, there are three other types of multiplier (Archer 1977b).

1. *The sales or transactions multiplier*. This measures the effect of an extra unit of tourist expenditure on activity levels in the economy, i.e. the increase in business turnover created by the tourist expenditure;
2. *The output multiplier*. This relates a unit of extra tourist expenditure to the increase in the level of output in the economy. The output multiplier differs from the sales multiplier in that, in addition to sales generation, the output multiplier also includes changes in the level of inventories;
3. *The employment multiplier*. This is the ratio of direct and secondary employment generated by additional tourism expenditure, to direct employment alone. Occasionally it is calculated as the amount of employment generated per unit of tourist expenditure.

Each of these types of multiplier measures a different phenomenon. Each has its own utility. For example, if one is comparing the effects of a number of projects one may be interested in their implications for both income and employment. Depending upon the priorities, and the flexibility of investment the one with the highest income or employment multiplier, or the best combination of the two, could be chosen.

To add to the complexity and confusion, income multipliers are calculated in a number of different ways in the literature, with implications for results. Assuming that the injection of one dollar of tourist spending creates 50 cents of direct income and 25 cents of indirect and induced income, then the variations in calculations are as follows:

1. Income multipliers may be calculated as the ratio of the total income generated (primary, indirect and induced) to the direct income created by the primary expenditure:

$$\text{i.e. Multiplier} = \frac{\text{primary + secondary}}{\text{primary}} \text{ (i.e. indirect and induced)}$$
$$= \frac{0.5 + 0.25}{0.5}$$
$$= 1.5$$

   This is known as an 'orthodox' income multiplier;

2. Income multipliers may be calculated as the ratio of the total income generated to the total initial expenditure:

$$\text{i.e. Multiplier} = \frac{\text{primary + secondary}}{\text{initial expenditure}}$$
$$= \frac{0.5 + 0.25}{1.0}$$
$$= 0.75$$

   This is known as an 'unorthodox' tourist multiplier.

Although the data used in each case are the same, quite different multiplier coefficients emerge with variations in the methods of calculation. It is imperative that interpretations of multiplier coefficients take into account the method of calculation, and that comparisons are only made between coefficients which have been calculated in the same way, if accurate policy information is to be derived.

Archer (1976: 119) believed that the principal weakness of the multiplier concept is not so much the model itself but, rather, the way it has been used and interpreted. One of the most notorious examples of the misuse of tourist multipliers is contained in the *Zinder Report on the Future of Tourism in the Caribbean*. Copies of this report are exceedingly difficult to locate. However, two excellent critiques of the report are readily available (Levitt and Gulati 1970; Bryden and Faber 1971), and their comments have been used to illuminate this discussion.

The Zinder Report claimed that for all nine territories of the Eastern Caribbean, tourist expenditures always yielded 2.3 times the initial expenditure as a result of successive rounds of transactions. The application of 2.3 to all nine territories of divergent characteristics, alone, should be enough to raise doubts concerning the accuracy of the study. However, in addition to the 'slipshod' approach employed in arriving at the coefficient of 2.3 (Levitt and Gulati 1970: 328), the report made insufficient allowance for leakages and only considered the value of transactions. One of the major criticisms of the report concerned the method employed to calculate the multiplier. The calculations began with an estimate that $1,000 of tourist expenditure could be broken down into four categories:

$315 spent on accommodation;
$385 spent on food and beverages;
$150 spent on other purchases;
$150 spent on sightseeing.

Each of these four categories of expenditure were taken through four rounds of spending. It was assumed that by tracing these rounds for each category of expenditure one could obtain a close approximation of total national income generated by tourism. The Zinder calculation for accommodation was as follows:

First Round	Tourist pays to hotelier	$315.00
Second Round	Hotelier pays wages, salaries, taxes, and local goods and services	$248.37
Third Round	Wage and salary earners purchase goods and services; local suppliers pay wages and purchase other goods and services; Government spends tax revenue in the local economy	$151.91
Fourth Round	Same as above for third round	$ 61.00
Estimated Total Spending		$776.28

The calculation of the estimated total spending figure assumed that the $61 did not completely disappear but that it had a further multiplier effect through additional rounds of spending. The estimated annual turnover per dollar is arrived at by dividing the estimated total spending by the initial tourist expenditure ($776.28 ÷ $315.00 = 2.46) and the result indicates that the $315 'turns over' 2.46 times in a year. A similar procedure was then implemented for the food and beverage category which was calculated to be 2.22. The accommodation coefficient was then applied to sightseeing, and the beverage-food figure to purchases. A weighted average of 2.3 was derived for the four expenditure categories. This, the report concluded, was the value of the tourist multiplier.

In addition to the general comments mentioned above, Bryden and Faber (1971) and Levitt and Gulati (1970) had more specific criticisms:

1. Income generated by expenditures on investments in the expansion of hotels and similar developments were omitted;
2. Successive gross receipts were summed, rather than the successive incomes received by residents;
3. There was a failure to consider the propensity to import as a major contributor to leakages. In failing to allow for import leakages, the nominal values of a number of transactions were added together. This, according to Bryden, 'multiplied the multiplier'. Bryden and Faber carried out their own analysis and obtained an income multiplier of between 0.6 and 0.76 for the island of Antigua. They claimed: 'What can be asserted

> strongly . . ., is that the relevant tourist multiplier is extremely unlikely to be as high as 1.00 in most East Caribbean Islands, and is certainly nowhere near the value of 2.3 claimed for it by the Zinder Report (Bryden and Faber 1971: 69).

> The Zinder Report is but one of a number of examples where multipliers have been calculated incorrectly and interpretations have been misleading.

Applications of tourist multipliers

The magnitude of income multipliers varies considerably between national and regional destinations. National multipliers are usually higher than those calculated for regions, because of the larger economic base and the proportionately smaller leakages. Archer (1977b) listed documented estimates of tourist multipliers as follows:

Antigua	0.88
Bermuda	1.09
Dominica	1.19
United Kingdom	1.68 to 1.78
Ireland	1.77 to 1.90

However, national tourist multipliers are an inaccurate guide to regional experiences because of the high leakages out of local economies. It should be self-evident that countries are more self-contained, in terms of domestic products and services, than regions and, therefore, experience a greater proportion of tourist expenditures as income.

A number of fairly sophisticated multiplier models have been used to calculate incomes generated by tourism at local and regional levels. These in-

Table 7 Local income coefficients – Edinburgh

Type of accommodation used by tourist	Direct	Indirect	Induced	Total
Hotel	0.205	0.045	0.043	0.293
Guest-house	0.210	0.071	0.049	0.330
Bed and breakfast	0.143	0.077	0.038	0.258
Tent	0.177	0.031	0.036	0.244
Caravan	0.161	0.028	0.033	0.222
Friends and relatives	0.143	0.031	0.030	0.204
Halls of residence	0.279	0.032	0.054	0.365
Other	0.233	0.031	0.046	0.310
Day trips	0.144	0.029	0.030	0.203
Staying visitors (weighted average)	0.191	0.050	0.042	0.282
All visitors (weighted average)	0.189	0.049	0.041	0.279

Source: Vaughan 1977a: 21

clude studies in Gwynedd, North Wales (Archer, Shea and Vane 1974), the Isle of Skye (Brownrigg and Greig 1975), Greater Tayside (Henderson, 1975), and the Lothian Region around Edinburgh in Scotland (Vaughan, 1977a). The case studies of Gwynedd and the Lothian Region will be employed to illustrate the meanings of multiplier coefficients, and their variations between different categories of tourists as indicated through the types of accommodation used.

Vaughan found that the income multiplier in Edinburgh varied between 0.203 for day trippers to a high of 0.365 for those accommodated in halls of residence designed for students (Table 7). The coefficients indicate that hotel users spend their money in such a way that more income is generated per unit of tourist expenditure than for those using caravans, tents or staying with friends or relatives. The table also illustrates the components of local income multipliers. The largest contribution to local income was made to the direct recipients of tourist expenditures, with progressively less being generated at subsequent phases of spending.

Table 8 Tourist multipliers in Gwynedd, 1973

Category of tourist (by type of accommodation used)	Direct income (1)	Direct, indirect and induced income (2)	'Orthodox' income multiplier [(2) ÷ (1)] (3)	'Unorthodox' income multiplier (4)
1. Hotels and guest-houses	0.2265	0.3237	1.43	0.3237
2. Bed and breakfast	0.5775	0.6351	1.10	0.6351
3. Furnished accommodation	0.3389	0.4012	1.18	0.4012
4. Caravans	0.1407	0.2097	1.49	0.2097
5. Holiday cottages	0.1786	0.2682	1.50	0.2682
6. Tents	0.1937	0.2845	1.47	0.2845
7. Friends and relatives	0.1767	0.2836	1.60	0.2836
8. Day trippers	0.2113	0.3032	1.43	0.3032
9. Others*	—	—	—	—
Composite multiplier*	—	—	1.34	0.3682

* Multipliers for holiday camps, hostels, etc, are more difficult to assess and have been omitted from this table.

Source: After Archer, Shea and Vane 1974: 42

Similar results were found by Archer, Shea and Vane (1974) in their study of Gwynedd. The income multipliers for each of eight categories of tourists, together with a composite tourist multiplier, are shown in Table 8. The first column shows the direct income generated in Gwynedd by £1 of tourist expenditure. As in the Lothian example, income generated by bed and breakfast lodgers (58 pence in £1 of tourist expenditure) is proportionately higher than that of tourists staying in caravans (14 pence in £1 of tourist expenditure). In the same example, the additional secondary income generated by £1 of expenditure by caravanners is 7 pence (0.2097 minus the direct income effect of 0.1407), whereas the additional secondary income attributable to £1 of expenditure by bed and breakfast users is 6 pence (0.6351–0.5775).

Table 8 also illustrates the differences in the multiplier coefficients with the variations in methods of calculation. In column three the 'orthodox' income multiplier (the ratio of direct, indirect and induced income to direct income alone) has a composite figure of 1.34. Archer, Shea and Vane (1974: 41) concluded that this measure is of little importance since it reflects intra-regional production and consumption relationships, rather than measuring the effect of tourist expenditure. In the case of 'orthodox' multipliers, the coefficient will be greater than unity unless the indirect and induced income is negative. The 'unorthodox' income multiplier (the ratio of the total income generated to the total initial expenditure) of 0.3682 provides a more realistic picture of the combined effects of tourist expenditures as in small areas the coefficient is likely to be less than unity because of high leakages.

Table 9 Income multipliers of Gwynedd, Lothian and Skye

Category of tourist	Gwynedd	Lothian	Isle of Skye*
Hotels	0.3237†	0.2930	0.3300
Guest-houses	—‡	0.3300	0.3900
Bed and breakfast	0.6351	0.2580	0.3900
Tent	0.2845	0.2440	0.2700
Caravan	0.2097	0.2220	0.3000
Friends and relatives	0.2836	0.2040	0.2600
Halls of residence	—‡	0.3650	0.2700
Holiday cottages	0.2682	—‡	0.2850
Day trippers	0.3032	0.2030	0.2200
Composite multiplier	0.3682	0.2648	0.3016

* Composite coefficients for Isle of Skye were calculated as upper and lower estimates only were provided.
† Guest-houses were included in this calculation and, hence, there is no single value for this category.
‡ No value was given for this category of tourist.
Sources: Vaughan 1977a; Brownrigg and Greig 1975; and Archer, Shea and Vane 1974.

In Table 9, the 'unorthodox' multipliers of the Gwynedd study and the total income multipliers of the Lothian study are compared. It can be seen that the values exhibit considerable similarity. An additional column shows the results of another British study undertaken in the Isle of Skye (Brownrigg and Greig 1975: 267). There are some variations in the coefficients for individual, specific tourist categories, for example the relatively high coefficient for Gwynedd bed and breakfast, but the findings are generally consistent across the studies.

Canadian values for tourist income multipliers are similar to those calculated in the United Kingdom. In the Muskoka region of Ontario, Triantis (1979: 276) concluded that $1 of tourist expenditure results in 26.5 cents of direct income for local residents. These 26.5 cents are respent to yield an additional 5.7 cents of income in the second round of spending, 1.2 cents in the third round and 0.3 cents in the fourth. Income generated from one dollar of tourist expenditure is approximately 34 cents and, hence, the 'unorthodox' income multiplier is 0.34.

Multiplier models vary greatly in their sophistication and rigour. Simplified, 'ad hoc' models can be expected to yield less accurate results than those calculated by input-output analysis. Input-output tables show the flows of current transactions through an economy for a specified period of time, usually a year. The various types of business activity are grouped into sectors. Tables show the total value of sales made by each sector to the other sectors and the purchases made by each sector from each of the other sectors (Archer 1977c: 1). Input-output analysis facilitates the conversion of these data into information on income generation. This technique has been used by Sadler and Archer (1974), Diamond (1976) and Anthony (1977).

Regardless of the sophistication of the multiplier models which are used, the accuracy of results ultimately depends upon the adequacy of the data. According to Bryden (1973), Archer (1977b) and others, this is one of the major problems of multiplier models. The problem is particularly serious at the regional and local level because, at such scales, data are either not available or not in the form required for rigorous multiplier analysis. Part of the problem is a reflection of the nature of the tourist industry itself. It is a multiproduct industry which encompasses a number of different economic sectors. Tourism expenditure is spread across a considerable range of activities and, hence, patterns of expenditure are often difficult to determine.

Income multipliers are useful for assessing the short-run economic impacts of tourist expenditure. On a short-term basis, multipliers can provide a wealth of information about tourist impacts. They can help to:

1. Identify weak linkages in the economy;
2. Provide information on the degree to which such objectives as maximizing income and employment, and minimizing foreign exchange losses, are being met;
3. Identify areas in the economy which require stimulation, and others which bring large benefits and merit expansion.

The major value of multipliers is their utility in short-term economic planning. Since an assumption of the employment of multipliers is that the structure of the economy remains unchanged, their value for long-term prediction is more limited.

Archer has written extensively on tourist multipliers, their application and limitations. The selection of his references which are included in the bibliography provide an excellent introduction to these topics.

Income generation and the distribution of tourist spending

Following the consideration of multiplier coefficients, it is appropriate to examine two further topics related to income:

1. The total income generated for residents of a destination;
2. The distribution of that income among residents of the destination, i.e. who benefits from tourist spending?

Total income generation

Again, it is useful to examine the United Kingdom examples of Gwynedd and the Lothian Region, as a basis for discussion. As differences between the multiplier coefficients were usually found to be quite small between different tourist types, they were not major factors altering the proportion of total income generated. However, this need not always be the case.

Table 10 Total income generation for Edinburgh

Type of accommodation used by tourist	Expenditure		Total income		Income as per cent of expenditure for each tourist type
	(£000s)	(%)	(£000s)	(%)	
Hotel	7,738	42.5	2,267	44.7	29.2
Guest-house	3,245	17.8	1,070	21.1	32.9
Bed and breakfast	1,926	10.6	497	9.8	25.8
Tent	437	2.4	106	2.1	24.2
Caravan	619	3.4	138	2.7	22.2
Friends and relatives	2,452	13.5	499	9.8	20.3
Halls of residence	560	3.2	204	4.0	36.4
Others	410	2.2	127	2.4	30.9
Day trippers	804	4.4	163	3.2	20.2
Total	18,191	100.0	5,072	100.0	

Source: After Vaughan 1977a: 28

In the Lothian Region, the £18.2 million of tourist expenditure resulted in approximately one third that (£5.1 million) in income for Edinburgh residents (Table 10). This smaller amount of local income, as compared with the total injection of money, is a reflection of the inability of any local economy to extract more than a fraction of any monetary injection because of leakages such as taxation and the purchase of goods from outside the area. As the multipliers are not greatly dissimilar in this case, the most important factor determining the impacts of different types of tourists is the size of their total expenditure. As may be expected, since hotels claimed the largest proportion of total visitor expenditures (42.5 per cent), they also contributed the most to local income (44.7 per cent). Tourists staying with friends or relatives spent 13.5 per cent of all expenditures but, with a relatively small multiplier, they generated only 9.8 per cent of all local income from tourism. The last column of Table 10 indicates the efficiency with which expenditure is converted into income by each accommodation sector. Guest-houses and halls of residence,

Table 11 Total income and output generated within Gwynedd by tourist expenditure: 1973 (June–Sept.)

Category of tourist (by type of accommodation used)	Expenditure (£000s)	'Unorthodox' income multiplier	Income created (£000s)	Output created (£000s)
Hotels and guest-houses	10,867	0.3237	3,518	15,357
Bed and breakfast	7,870	0.6351	4,998	6,310
Furnished accommodation	7,593	0.4021	3,046	7,955
Caravans	6,995	0.2097	1,467	7,694
Holiday cottages*	1,008	0.2682	270	1,398
Tents	2,990	0.2845	851	4,204
Friends and relatives	1,204	0.2836	341	1,518
Holiday camps†	657	—	138	723
Day trippers	1,022	0.3032	310	1,431
Others†	861	—	181	947
Total	41,067	0.3682	15,120	47,537

* This does not include the payment of rates to local authorities nor the expenditure of money on house improvements or extensions.

† The multipliers for holiday camps and 'others' are difficult to assess and their respective income and output figures are intended to illustrate the most likely degree of magnitude.

Source: After Archer, Shea and Vane 1974: 48

in particular, generate more income per unit of expenditure than other types of accommodation. In excess of 36 per cent of total expenditures of those staying at halls of residence is returned as income, compared with only 20 per cent for those staying with friends and relatives.

In Gwynedd the total income generated was approximately 37 per cent of total tourist expenditure. Table 11 summarizes the overall monetary impact of tourist expenditures in Gwynedd and also indicates the part played by each accommodation sector. Expenditures of visitors staying in hotels and guest-houses made up just over one quarter of all tourist expenditures but created just under a quarter of the total income. Bed and breakfast accommodation only attracted approximately one fifth of total expenditure but generated approximately one third of the total income. The proportionately high income generated by bed and breakfast patrons is a result of relatively low leakages. Hotels and guest-houses have to pay for goods, services and capital equipment purchases from outside of Gwynedd and this, together with payments made to seasonal, 'imported' labour, increases the leakage and reduces the generation of income.

The total income generated by tourist expenditures increases as one extends the area in which measurement takes place. For example, the £18.2 million of expenditure in Edinburgh generated approximately £5 million for the city of Edinburgh, a further £66,000 in the rest of the Lothian region, and an additional £793,000 in the rest of Scotland and elsewhere (Vaughan 1977a: 31). As the area of investigation increases, the higher the income multiplier or the proportion of expenditure which is generated as income. This is due to the reduction in the leakage which, in turn, reflects the greater ability of the larger region to supply the needs of the industry from within its boundaries.

The beneficiaries of tourist spending

The tourist industry consists of an heterogeneous group of establishments which provide a wide range of goods and services for tourist consumption (Archer 1972: 42). Some of these firms are totally dependent on tourist spending while others also cater to local residents and revenue from tourists forms only a small proportion of their business. Many different sectors of the economy, therefore, will be influenced by tourist spending. The economic impact of tourist spending depends upon the distribution of the initial round of tourist expenditures and the linkages within the economy.

Many studies discuss the distribution of the initial round of tourist spending but relatively few examine the composition of secondary flows and determine which sectors of the economy benefit from the multiplier effect. Figure 4 shows how the tourist dollar was spent in Florida in 1968 (Lundberg 1972: 139). Comparison of these with other findings, even in the same continent, are very difficult because the categories of expenditure employed in the various studies differ. Furthermore, destinations are not influenced by the

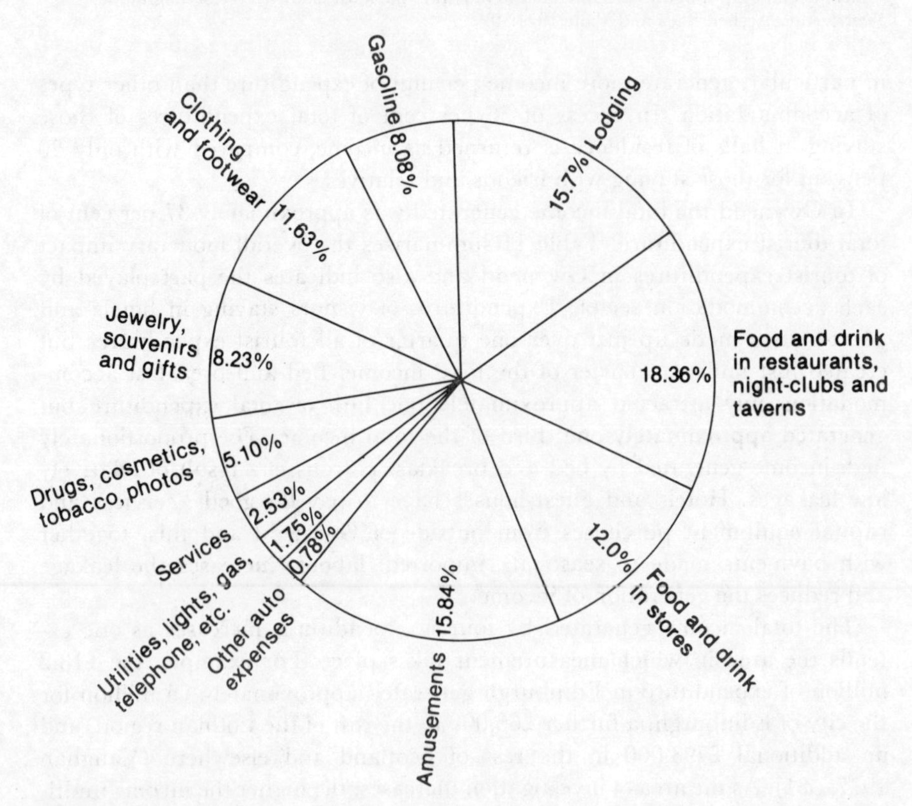

Fig. 4 Distribution of the tourist dollar in Florida 1968 (*Source*: Lundberg 1972: 139)

same economic variables from one year to the next. For example, it is difficult to compare a study done in 1960 with one done a decade later as disposable incomes and price differentials are radically different. These have a considerable bearing on travel behaviour.

As Archer (1972: 42) has suggested, it would be helpful if consistent categories were adopted by researchers for this would facilitate the comparison of case studies. He proposed a four-category classification: food, lodgings, transport and other purchases. Little (1962: 30) and Clawson and Knetsch (1966: 235) both used these groupings in their studies in the United States. They concluded, with some degree of consistency, that food took between 25 and 35 per cent of tourist expenditure, lodgings 22 to 27 per cent, and transport 20 to 23 per cent. Other purchases varied between 17 and 27 per cent.

There are few analyses of indirect and induced regional money flows and their beneficiaries. These secondary effects are most clearly revealed by input-output analysis. Harmston (1969: 9) examined both the direct and indirect impacts of tourist expenditure in Missouri and showed that the main secondary beneficiaries were in real estate, rentals and retailing. A more recent study by Anthony (1977) measured the effects of the 1975 expenditures of Americans travelling in the United States. He found that the direct effects of tourist expenditure were most prominent in retail trade, airlines, accommodation, entertainment and auto services and repair. Indirect effects were greatest in real estate, wholesaling, maintenance and construction, trucking and warehousing. Retailing, although it was highest in direct effects, ranked low in indirect and induced effects. The main recipients of tourist spending, particularly secondary spending, are service industries. In Anthony's analysis, the top five categories (retail, airlines, accommodation, motion pictures and amusements, and auto services and repair) received nearly 65 cents of every dollar of domestic travellers' expenditures. Apart from benefits to construction, there appear to be few linkages to manufacturing and agriculture or to other sectors of the economy.

Revenues for governments

Income from tourism for governments arises from direct taxation, mainly on tourist manpower, tourism and transport enterprises, fees and service charges; from indirect taxation from customs duties and on goods consumed by tourists; and from interest payments, loan repayments and revenue from government-owned or financed tourist enterprises. The major governmental income earners are usually direct taxation and customs duties. Occasionally governments have tried to estimate the direct tax revenues from tourism, for example, from hotel and airport taxes. According to Clement (1967: 71), this is a static approach which leads to distorted results. What is really important is the total amount of tax revenues, both direct and indirect, that would not accrue to the government if there were no tourism.

The total revenue generated for the government of the Bahamas by international tourism in 1974 is shown in Table 12. Tourism contributed more than B$45 million of customs duties which was 64.3 per cent of all revenues

Table 12 Government revenue from international tourism, Bahamas, 1974 (B$)

Head number	Item	Revenue from tourism	Total government revenue from all sources	Tourism revenue as a percentage of the total head number
1.	Customs duties	45,637,000	70,989,675	64.29
2.	Excise duties	—	3,141,108	0
3.	Real property tax	2,064,000	2,839,992	72.67
4.	Motor vehicle tax	2,085,000*	2,880,850	72.37
5.	Gaming taxes	8,816,000	8,898,413	99.07
6.	Tourism tax	4,849,000	4,955,740	97.85
7.	Stamp tax	—	6,473,742	0
8.	Other taxes	1,544,000*	6,054,447	25.50
9.	Fees and service charges	3,494,000*	5,719,926	61.08
10.	Revenue from Government property	—	2,411,975	0
11.	Interest	—	458,193	0
12.	Reimbursement and loan repayment	—	5,881,041	0
13.	Services of a commercial nature	6,077,000*	7,547,878	80.51
	Total	74,566,000	128,252,980	58.14

Note: Total government revenue figures for 1974 were obtained from the Draft Estimates of Revenue and Expenditure 1976.
* Less confidence should be placed in the values of the figures marked with an asterisk.
Source: Archer 1977a: 40

in that category. The second largest item was the gaming tax, which was principally composed of casino taxes, and these amounted to B$8.8 million. Tourism also contributed high proportions of total government income obtained in real property taxes, motor vehicle taxes, fees and service charges, and commercial services. Fees and service revenue was obtained largely from the inspection fees earned by the Customs and Immigration Departments.

Research which examines the overall contribution of tourism to the tax base or to the incomes of governments is limited. Information of this nature would be invaluable, particularly for developing countries attempting to derive urgently needed income from the development of tourism.

Tourism and employment

As has already been demonstrated, a considerable quantity of research has been undertaken on the effects on income of tourist expenditures. However, less is known about the impact of tourist expenditures on the creation of employment opportunities. McCloy (1975: 49) posed a number of questions on relationships between tourism and employment for which researchers must find answers:

1. How many people are employed as a result of the travel industry?
2. What types of job opportunities are available in the tourist industry?
3. What skills do people require and what returns and benefits can be expected from their employment?
4. What is the geographical distribution of this employment?
5. What capital investment is required to create this employment?
6. What is the overall economic contribution to national, regional and local economies of this employment?
7. What will be the future significance of the travel industry as a generator of employment?

Current research has yet to provide complete answers to all of these questions, although recent studies have thrown considerable light on the more specific ones (1, 2 and 5). In addition, new methods are being developed for the compilation of more accurate and comprehensive tourist employment statistics. There is general agreement that tourism creates employment and that this is usually a beneficial effect. There are employment opportunities in hotels, and tourists spending money outside of their hotels create additional income earning opportunities for taxi drivers, curio shops, restaurants, travel agencies and entertainment facilities. The building of hotels and the upgrading of transportation, sanitation and water supply facilities provides employment in construction. Increased demands for food could also increase the number of jobs in agriculture.

Three types of employment are generated by tourism and they closely correspond with the types of income discussed in the section on multipliers (Goffe 1975: 26):

1. Direct employment results from visitor expenditures in tourist plants, such as hotels;
2. Indirect employment is still in the tourist supply sector but does not result directly from tourist expenditures;
3. Induced employment is the additional employment resulting from the effects of the tourism multiplier as local residents respend the additional money which they have earned.

Most assessments of jobs created by tourism make reference only to direct employment. Ouma (1970: 103), for example, noted that the following numbers of people were directly employed in tourism in Kenya: 7,500 in hotels and lodges; 1,300 as tour operators, organizers, travel agents and in car hire; 500 safari outfitters; 200 in air charter; 1,500 in curio and other sales; and 1,600 by government and national parks. Wall and Maccum Ali (1977: 45) indicated that, in Trinidad and Tobago, the creation of two hotel rooms generates permanent employment for three persons. They cited the example of the new Hilton Hotel in Port of Spain generating 400 jobs.

More detailed investigations have been undertaken by Jud (1974), Elkan (1975) and Frechtling (1975). Elkan's study in Kenya and Tanzania elicited information about the numbers employed in different occupations, the total wages bill, and the numbers of non-Africans engaged in the industry. Tourism in these two countries employed 11,000 people. This may not seem to be very many when it is pointed out that their combined enumerated population is 1.1 million. However, when compared to employment figures, for example the 113,200 employed in manufacturing in 1972, tourism is clearly an important provider of jobs. All of the above studies have been limited to an examination of direct employment.

The employment of tourist multipliers, coupled with the increasing research emphasis of government tourist organizations, has prompted a number of recent studies of secondary employment effects. Archer (1973: 7), using an adaptation of a tourist income multiplier model, calculated the employment generated by tourist activity in Anglesey (Gwynedd). He calculated that for every nine jobs created directly by tourism spending, an additional one job, or job equivalent, also resulted. It was also calculated that £20,930 of tourism expenditure at 1970 prices was required to generate those nine jobs. Archer (1973: 76) also found that while tourist spending generated approximately the same income as the equivalent amount of general spend-

Table 13 Edinburgh: Total employment multiplier at the city level (per £1,000 of visitor expenditure)

Serviced accommodation		Unserviced accommodation	
Hotel	0.187	Tent	0.121
Guest-house	0.208	Caravan	0.113
Bed and breakfast	0.284	Friends and relatives	0.126
Halls of residence	0.331	Others	0.250
Weighted average	0.210	*Weighted average*	0.136

Source: After Vaughan 1977a: 34.

Table 14 Edinburgh: Components of the employment coefficients at the city level (per £1,000 of visitor expenditure)

Type of accommodation used by tourist	Direct	Indirect	Induced	Total
Serviced accommodation	0.157	0.025	0.029	0.212
Unserviced accommodation	0.101	0.013	0.022	0.136
Weighted average	0.142	0.022	0.027	0.192

Source: Vaughan 1977a: 35

ing, the employment created per unit of tourist expenditure was more than twice that created by general spending. It was noted that 2.39 jobs were created by each additional £10,000 of general spending, compared with 4.83 jobs in the case of tourism.

Vaughan (1977a: 10) also used employment multipliers with similar results to those achieved in the Anglesey study. For example, for halls of residence, the total employment multiplier was 0.331 (i.e. £10,000 of tourist expenditure created 3.31 jobs). The figure for hotels was 0.187 and for caravans it was as low as 0.113 (Table 13). In other words, visitors using the most intensive types of accommodation generate the largest number of employment equivalents. Vaughan's study is also instructive in that he divided the total employment created into direct, indirect and induced categories and calculated the proportion of the employment multiplier attributable to each (Table 14). Not surprisingly, more jobs were created in the direct employment category.

Other studies, using less sophisticated methodologies, include total employment figures but most do not relate them to a given amount of tourist expenditure. For example, Bond and Ladman (1973) indicated that 90,000 additional jobs were created in Arizona in 1972 as a spin-off of 'basic' employment in tourism. This figure was calculated by the application of a multiplier of 2.6 jobs in 'non-basic' activities for every position in 'basic' employment.

The studies employ different terminologies, categories and methodologies and this makes comparison difficult. There is a paucity of detailed empirical studies on tourism and employment. Nevertheless, sufficient evidence is available to make some general observations. Firstly, employment and income effects are closely, though not perfectly, related. They are analogous in that primary or direct employment and income can be distinguished from secondary or indirect employment and income. There is a causal relationship between tourism-generated income and employment but they are not necessarily of equal size nor need they be created at the same time. Income and employment multipliers are not identical and the maximization of income does not necessarily lead to the simultaneous maximization of employment.

Secondly, the effects on employment are influenced by the type of tourist activity. Some types of tourism are more labour-intensive than others. Accommodation facilities, particularly hotels and boarding-houses, employ a substantial proportion of the labour force in destination areas. Elkan (1975: 125) noted, for example, that hotels and motels in East Africa employed 0.8 persons per hotel bed. Jud (1974: 32) provided similar figures for Mexico and indicated that employment in the hospitality activities had grown more rapidly than total employment in tourism or in other components of the service sector. Between 1960 and 1970 employment in hotels and motels increased by 111 per cent while the growth in employment in restaurants, cafes and bars was 147 per cent. Table 15 provides Canadian evidence that the majority of the tourist labour force is employed in the service activities of accommodation, restaurants and entertainments, and retailing. Most service industries are labour intensive. Nevertheless, although hotels,

Table 15 Employment by sector in the tourist industry in Prince Edward Island, July/August 1975

Tourist sector	Total number of employees	% of Total
Accommodation	2379	14.2
Restaurants – entertainment	2085	12.5
Processors	1753	10.5
Tourist attractions	1018	6.1
Gas stations – garages	1082	6.5
Transportation	525	3.1
Wholesalers	563	3.3
Rental operators	106	0.6
Retail dry goods	2494	14.9
Handicrafts	223	1.3
Grocery stores	1106	6.6
Construction contractors	2990	17.9
Other	353	2.5
Total	16,677	100.0

Source: Prince Edward Island, Department of Tourism, Parks and Conservation 1975: 18

motels and restaurants employ large numbers of people, they also require large amounts of capital. Jud (1974: 33) estimated that every job created in an hotel or motel in Mexico in 1960 required a capital outlay or more than $3,590. This compared with only $2,780 in manufacturing industry. It might be noted that Jud's finding contrasts with that of Bond and Ladman (1972: 46) which was reported earlier in this chapter. The financial resources required to create employment also vary with hotel size, with a greater running cost per bed in larger hotels.

A third generalization is that effects on employment are influenced by the types of skills which are available locally. It has been mentioned many times that tourism requires large numbers of workers with minimal skills. Tourism creates few jobs at a managerial and professional level and these are often filled from other sectors of the economy or by people brought in from outside the region. A number of studies show this trend: e.g. a New Brunswick (1972) employment study revealed that of the 80 per cent of employees in the tourist labour force in the accommodation and food services categories, only 4 per cent were in managerial or professional occupations. As many jobs are seasonal and involve janitorial work, housekeeping, and food and beverage preparation, it is not surprising to find that women employees outnumber males by at least three to one. The employment structure and the cyclical pattern of employment are often cited as disadvantages of the tourist industry. Young (1973: 115) stated that the low productivity potential of work in the tourist industry can have a depressing effect on regional economic growth.

A fourth consideration is that the employment effects of tourism are often disguised. In creating employment, tourism may take people from other sectors of the economy, particularly rural people; people not normally considered as part of the available work-force, such as married mothers and the retired; the unemployed; and individuals from work-forces outside the monetary economy. This makes it difficult to gauge the real impact of tourism on employment. Impacts of tourism reach beyond a narrowly-defined destination area so that a concentration upon the destination area alone can mask other employment changes. The new jobs are often part-time. They may be filled by people taking a second job, in which case tourism induces a fuller utilization of those who are already employed but may not lead to appreciable decreases in unemployment figures. In other cases, individuals may leave another industry to take employment in tourism. This will result in vacancies in other sectors of the economy and these may be filled by imported labour. The importation of labour may have negative effects because it usually intensifies leakages. Young (1973: 166) cited the example of a new 125-room Holiday Inn hotel in the Caribbean which required 200 staff for its operation. Only 60 were available locally and the remainder had to be imported. Thus the local employment structure prior to the advent of tourist development is as much a determinant of employment effects as tourism itself.

Fifthly, much employment in tourism is highly seasonal and contributes to fluctuations in the levels of local and regional employment. For example, in Anglesey 800 additional workers must be employed to meet the summer demand (Archer 1973: 9). This represents a 5 per cent increase in the labour force of the island. This is attractive to those who only require seasonal employment, such as students and some housewives and part-time employees. On the other hand, seasonal employment has inherent disadvantages. It tends to attract those on the periphery of the labour force. D'Amore (1976: 33) believed that this discourages outward mobility to more productive employment. For example, seasonal employment may attract workers away from full-time employment which may discourage non-tourist industries and activities from locating in the area.

In summary, it is evident that tourism creates employment and, with some minor reservations, this is generally of benefit to the economy. Research on employment in tourism has concentrated on determining the number of jobs created by tourism, the sectors of the industry which offer employment opportunities and, to a lesser extent, capital–employment ratios. Most research has failed to examine the skills involved in tourist jobs, the returns which can be expected, and the role of this employment in local and regional economic growth. Nevertheless, considerable progress has been made in the establishment of methodologies for estimating tourist-generated employment. A number of studies use a standard ratio approach in which a coefficient is calculated, adapted and applied to various destinations (Bond and Ladman 1973; Minnesota Department of Economic development 1977). In the latter study it was calculated that $14,793 of tourist receipts generated one job. Minnesota's 1976 tourist receipts were in excess of $1.3 billion which supported almost 90,000 jobs. In spite of the empirical base, there are dangers in the blind application of standards, for ratios between employment and receipts or income will vary from place to place with differences in the existing employment structure and types of tourist activity. A number of more sophisticated techniques for measuring employment have emerged over the last decade but, apart from the increased use of multipliers, they have not received much use in studies of tourism. Detailed accounts of methodological developments in the estimation of employment generation can be found in the articles by Archer (1973: 67–77) and Frechtling (1975).

Tourism and entrepreneurial activity

Few studies make a convincing case for the existence of external economies arising from tourist development. However, there is little doubt that the tourist industry exhibits backward linkages and that external economies have emerged. For example, improvements to local and regional transportation networks, to water quality, sanitation facilities and garbage disposal may have been prompted by the tourist industry but benefit other sectors of the economy. The construction of an international airport may provide improved access to other areas for local residents and locally produced goods. Tourism may also benefit property owners through positive effects on real estate prices, although this may create difficulties for young locals who wish to purchase property. Many researchers have admitted the presence of such economies but few have paid specific attention to them or discussed the extent to which local entrepreneurial activity may be promoted by tourist development.

The extent to which the tourist sector can establish linkages with local entrepreneurs depends upon:

1. The types of suppliers and producers with which the industry's demands are linked;
2. The capacity of local suppliers to meet these demands;
3. The historical development of tourism in the destination area;
4. The type of tourist development.

Lundgren (1973) examined these factors in the Caribbean and his study is one of the few detailed investigations of tourism-related entrepreneurial activity. Lundgren attempted to determine the nature and magnitude of entrepreneurial activity generated by different forms and rates of hotel development. There are certain merits in concentrating upon accommodation:

1. It is a part of the tourist sector which absorbs a large proportion of tourist expenditure, i.e. 35 per cent of the tourist dollar in the Caribbean is spent on accommodation;
2. The accommodation sector is an important producer of goods and services and requires an efficient supply system. Its demands must be met by various other sectors of the economy, of which the supply of food is one of the most basic.

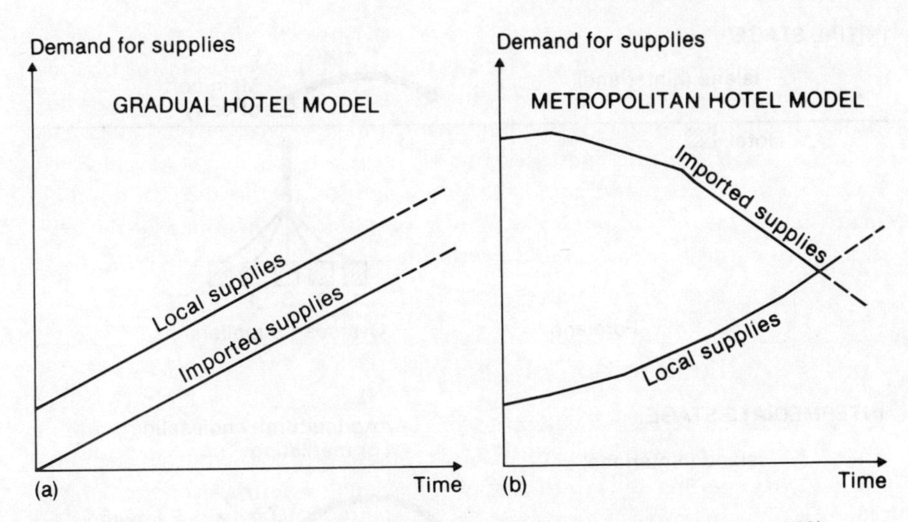

Fig. 5 Tourist-generated entrepreneurial activity (*Source*: After Lundgren 1973)

Demand-supply relationships between hotels and local entrepreneurs differ with the type of hotel development. Gradual hotel development is characterized by a succession of improvements and expansions of the infrastructure over an extended period of time. This creates a gradual increase in the demand for local produce. This is important to developing economies where there may be an inability to meet sudden increases in demand from local capacity. The early stage of this pattern of development is represented in Fig. 5(a). Initially, increased demand is paralleled by local supply. If demand continues to climb, the local supply capacity may be exceeded. This may be due to the restricted availability of space for agricultural expansion, a lack of technological innovation, and growing competition from foreign producers who are attracted by the expanding market. Eventually, because of inelasticity in the local food supply, there is a growing dependence on imported goods.

Most tourist development in developing countries emerges with large, metropolitan hotel complexes so that the pattern of entrepreneurship can be expected to be different from the one which has been outlined above (Fig. 5b). Hotel development in many of these countries has been rapid and has created an immediate demand for large volumes of agricultural products. Local suppliers are often unable to meet these demands. The hotels and foreign suppliers develop a closely integrated system which does not permit local suppliers to take advantage of the expanded market. Many developing countries have failed to progress past this stage of development and this has resulted in resentment and charges of neo-colonial domination.

Lundgren suggested that if the situation were permitted to change, two further stages of entrepreneurial development would be possible (Fig.6). Following the initial stage, which has been described, the intermediate stage would see the development of a locally based and controlled supply system. A large wholesale/marketing distributor would facilitate the participation of

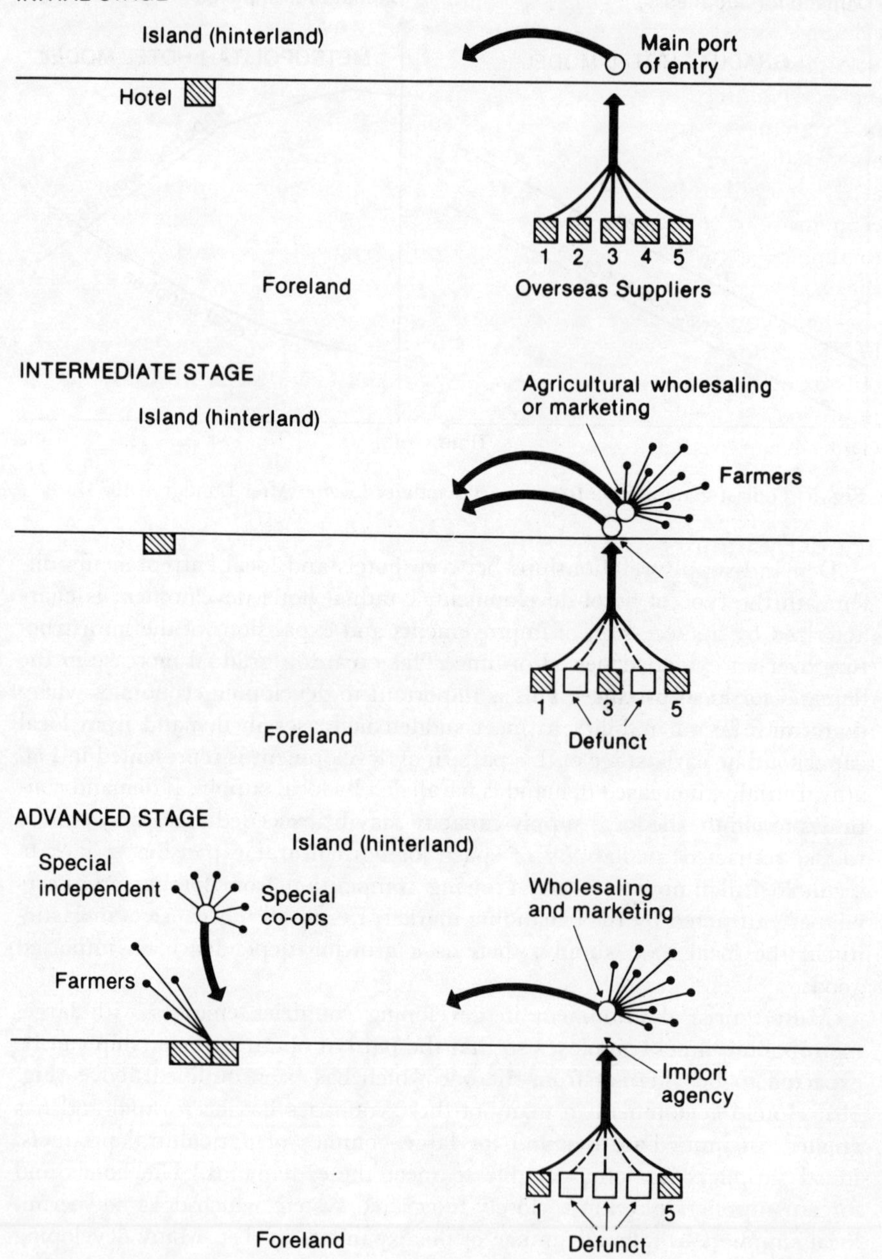

Fig. 6 Stages of entrepreneurial activity (*Source*: After Lundgren 1973)

local suppliers and this would result in the erosion of the foreign dominance of the supply system. The final, advanced stage sees further expansion of local wholesaling which, in turn, stimulates growth within agriculture, leading to an intensification of production and further hinterland development.

Lundgren's ideas are provocative but, at the present state of tourism development, they are best regarded as hypotheses which have yet to be tested.

Although it is attractive to think of a sequence of developmental stages, the exact pattern of entrepreneurial activity is likely to vary from place to place, in part because of the factors affecting linkages, which were listed above. Nevertheless, it is essential that the tourist industry is serviced, as far as possible, from local producers if its full potential contribution to the local economy is to be realized. This will depend, largely, on the ability of the destination area to break the stranglehold of foreign suppliers and to intensify the economic links with local suppliers.

The encouragement of entrepreneurial activity is generally regarded as a beneficial impact of tourist activity. The profits which result often accrue to only a small section of the local population. However, the indirect benefits of improved economic and social services have been sufficient to counter such criticisms among some observers.

Tourism and economic structure

Throughout this chapter it has been shown that tourism can often induce economic benefits for destination areas. It is likely that the development of tourism has been accompanied by other changes in the economic structure of destinations. However, little is known about the nature of such changes. The literature makes little reference to changes in industrial production as a result of tourist development. This may be because tourism and many types of industry are often viewed as being incompatible, so that one has expanded at the expense of the other. Nevertheless, it has also been assumed that industrialization is a necessary component of modernization and that tourism can accentuate that process. The greatest changes in economic structure have probably occurred when the transformation has been from an essentially primary producing economy to one dominated by tourism. As such changes are more pronounced and readily identifiable than less extreme modifications in economic structure, they have been mentioned in the literature, although coverage of these effects is far from exhaustive.

Changes in patterns of agricultural production in many rural economies are not endemic to tourism. Many of the changes have been the result of demographic pressures, technological progress, employment opportunities outside of the rural economy and modifications in patterns of land ownership. Tourism, though not always a major cause, has often contributed to the acceleration of such changes.

A principal change which has occurred in rural economies has been the occupational shifts of rural inhabitants. Many farmers and wage earners have left the land to pursue more lucrative jobs in the tourist industry or in construction. Jobs outside of agriculture may be more attractive so that few young people remain in rural areas and the future of farming in such areas is in jeopardy. The structural change from agriculture to tourism also creates changes in land use patterns. Tourism increases the competition for land, raising land prices and encouraging sales, contributing to the fragmentation of landholdings. Land is sold in smaller units and at higher prices and this contributes to inflation. The victims of this inflation are the young residents trying to purchase land or homes. At the same time the area becomes less self-sufficient and increasingly dependent on national and international economic conditions.

It is easy to paint such changes in a negative light. However, there are documented cases where the development of tourism has arrested, and even reversed, population decline in marginal agricultural areas (Diem 1980). Greenwood (1976: 138) has noted that it is possible for tourism to stimulate population growth and, simultaneously, lead to increased farm profits. As a result of the transformation, residents may have a higher standard of living but this may be gained at the expense of numerous social side effects which, depending on the perspective of the viewer, may be good or bad. These have been discussed by Friedl (1972), Greenwood (1976) and Pacione (1977) and also constitute the subject-matter of chapter five.

The economic costs of tourism

The emphasis of this chapter has been on the economic benefits which accrue to destination areas as a result of tourist development. This emphasis reflects the dominant orientation of research on the economics of tourism. Few studies have attempted to identify and describe the economic costs of tourism in a systematic fashion. Investigations have been limited largely to the measurement of the more obvious direct costs such as investments in facilities, promotion and advertising, transportation and other infrastructure. Most studies have failed to address the hidden, indirect costs, such as the importation of goods for tourists, inflation, the transfer of the profits of multinational corporations, economic dependence and opportunity costs.

Opportunity costs

Destination areas, in investing their scarce resources in the development of tourism, have seldom considered what the same resources could provide were they to be invested in another industry. In other words, it is necessary to investigate the relative economic benefits of investing in tourism as opposed to investing in another industry. This kind of comparison is commonly known as the 'opportunity cost' of an investment. Although Vaughan (1977c) has written a paper on the topic, there is little in the literature on opportunity costs. Parks Canada (1970), Lundgren (1973) and Jafari (1974) briefly mention the term but they provide little empirical evidence of the values of opportunities forgone.

In order to measure opportunity costs it is necessary to accurately describe and evaluate the opportunity or opportunities which are to be forgone. Parks Canada (1970) described the opportunity costs of a proposed park development in Newfoundland. Establishment of the park would mean the curtailment of all forestry operations which produce fuelwood, pulpwood and saw timber. From a national perspective the opportunity cost of establishing a park is small as it is unlikely that the timber harvest forgone in one area cannot be replaced by another harvest elsewhere in the country. From a regional or local point of view the opportunity cost may be considerably higher. A reduction of fuelwood may necessitate larger imports of heating oil, and reductions in pulpwood may increase the need to import newsprint. Parks Canada (1970: 34) estimated that establishment of the park would result in a net loss of $109,456 through replacement of lost fuelwood. This cost would have to be borne by residents of the local area.

Opportunity costs are extremely difficult to measure and this is a major reason why they are often ignored. However, the neglect of opportunity costs and the negative externalities of tourism have been partially responsible for an overemphasis on tourism as a positive factor in economic development. It is important that the obvious benefits do not blind decision-makers to the substantial costs.

Overdependence on tourism

Some destinations, by becoming overdependent on tourism for their livelihood, have made themselves vulnerable to changes in tourist demand. Although tourism is a growth industry and the total volume of tourist traffic is likely to increase in the foreseeable future, all destinations may not share in that growth. Tourism is highly susceptible to changes from within (e.g. price changes and changing fashions) and outside (e.g. global economic trends, political situations and energy availability) the industry. For example, political unrest at one destination can rapidly reduce demand for that location and, at the same time, divert it to others. Many tourists avoid destinations which are politically unstable, but they seldom cancel their travel plans completely and usually select an alternative. This transfer of demand can be disruptive for both locations. Greatly reduced patronage at one location means the underutilization of services, job redundancy and loss of income. The nature of the impact at newly selected destinations will depend upon their capacity to adapt and absorb the additional arrivals.

The religious confrontations in Ireland have been responsible for reductions in tourist arrivals there. The Arab–Israeli wars of 1967 and 1973 greatly affected tourist arrivals in the Middle East. Israel recorded 74,000 frontier arrivals in September 1973 but, with the advent of war, this slumped to 20,000 in the following month. The magnitude of the loss is influenced by the severity of the disruptive factors and the degree of dependence on the tourist industry. Israel and Ireland are both heavily dependent on international tourism for foreign earnings and the losses have been severe.

Some destinations have simply fallen out of fashion. Monaco, for example, had 70 hotels with 2,580 rooms in 1959, but thirty years later it had only 31 hotels and 1,650 rooms (Young 1975: 51). A major reason for the decline of Monaco was its inability to cater to the changing demands of tourists and the increasing diversity of tourist types. Monaco had catered primarily to the rich but, with the loss of this market and the advent of package tours, it was unable to provide facilities and services which were affordable to tourists with lower incomes.

Obviously a balance is required. Destination areas must ensure that visitor facilities grow in balance with the number of tourists. This requires accurate forecasting. However, even accurate forecasting would not completely solve these problems of destination areas for many aspects of the industry are beyond their control. Destination areas that are highly dependent on tourism are founded upon an unstable base. To avoid economic disruptions caused by fluctuations in demand, destinations should promote diversity both within the tourist industry and the base economy.

Inflation and land values

The inflationary consequences of tourism can arise in several different ways. Rich tourists can afford to buy items at high prices. Retailers, recognizing that their profit margins can be greatly increased by catering to tourists, increase their prices on existing products and provide more expensive goods and services. Such stores can compete successfully with those catering to local residents. They can afford to pay higher rents and taxes which are passed on to the consumer through higher prices. Local residents, in addition to paying more for their goods, may also have to go farther afield for their purchases as the diversity of local supply is reduced as stores catering to the local market are displaced by an increase in the establishment of specialty shops for tourists.

Inflation within destination areas is also caused by increasing land values. Growth in the tourist trade creates additional demand for land and competition from potential buyers forces the price of land to rise. The demand for more hotels, vacation homes and tourist facilities may bring sources of income to builders, real estate agents and landowners, but local residents are forced to pay more for their homes and larger taxes because of the increased land values.

The literature has not documented the extent to which inflationary price increases to residents are compensated by economic benefits, such as expanded employment and income. There have been attempts to measure the changes in land values caused by a variety of developments but there is no study of this type for tourism. There is also no empirical evidence on changes in the prices and availability of goods and services as a consequence of tourist development.

Seasonality

The seasonality of demand in most tourist regions is reflected in hotel occupancy rates. Many hotels actually close during the off season while others have greatly reduced revenues. Fixed costs make up a large proportion of total costs, so most hotels prefer to remain open all year round to secure as much revenue as possible. Nevertheless, production in the accommodation sector is greatly reduced in the off season. Since the investment is not fully used in the off season, the returns on capital are often low. This means that tourism is often a less attractive investment than other sectors of the economy which experience steady production. Hotels have also suffered from the development of other forms of tourist accommodation with lower capital inputs (Archer 1973: 13). The relatively low rates of return on much hotel investment have contributed to a shortage of hotel accommodation at peak periods.

Accommodation investments are not the only ones with a low rate of return. Tour operators also face similar problems. The reluctance of outside investors to become financially involved in marketing seasonal enterprises has meant that greater financial responsibility is borne by local investors. The opportunity costs of such investments are frequently high and other sectors of the economy may offer more attractive returns. Investment from public authorities may be necessary in the absence of interest from the private sector.

External costs

Tourist development imposes a number of other costs on residents of destination areas. These include:

1. Increased costs of garbage collection and disposal;
2. Increased maintenance costs for tourist attractions damaged by crowding and vandalism.

Again, there are few references to such costs in the literature.

Economic indicators of the future of tourism

The international economic system has provided an invigorating climate that has fostered the unprecedented growth of tourism in the past, and will continue to influence the destiny of the tourist industry. However, current economic conditions and predictions for the future appear more as obstacles to be overcome rather than reinforcers of growth. Wahab (1975: 170) has identified four limitations to future tourist development.

The first constraint is economic uncertainty stemming from inflation, fluctuating currency exchange rates, unemployment and sudden protective measures which are enforced on international money markets. The latter include revaluations of currency and the imposition of travel and departure taxes. Although inflation, for example, has diminished the proportion of incomes which is available for leisure purchases, the travel industry has managed to offset this by devising air charters, package tours, and concession fares, and has managed to hold travel costs at par with disposable incomes. However, it is doubtful if the industry can maintain its momentum given a proliferation of protectionist measures. Governments continue to employ policies designed to impede the flow of tourists out of their countries to conserve foreign exchange. For example, Mexico recently devalued its dollar and this drastically reduced the number of Mexican tourists visiting American resorts in the border states, and it also curtailed the spending of those who did elect to travel. It is apparent that the tourist industry is highly vulnerable to economic restrictions and to shifts in global economic trends.

A second constraint is the availability of energy resources at prices affordable to tourists and the operators of tourist services. Shortages of oil and escalating prices will undoubtedly affect the costs of travel but the full implications for the tourist industry are still unclear (Ditmars 1973; Waters 1974). The 1974 oil crisis was a particular problem for the aviation industry. In October 1973, American airlines were required to reduce their fuel consumption by 12 per cent, with a further 15 per cent reduction by December. This resulted in the cancellation of one in four daily departures, with some cities receiving 50 per cent reductions in the departure levels of the previous year. The economic loss through grounded aircraft and lost patronage was dramatic. The combined net profit for all world scheduled airlines was $41 million in 1974 compared with $434 million in the previous year. Most of this drop was attributable to the energy crisis. The effects on revenue and employment in destinations which suffered declines in tourist arrivals are less apparent and more difficult to measure.

The vulnerability of tourist development to changes in the availability and price of oil can also be seen in the rapidity of recovery once oil supplies were assured. Within months of the peak of the crisis, airlines were flying with 95 per cent of their fuel allocations and they were recording record profits by the middle of 1974. By 1976 patterns of world air travel had fully adjusted and annual increases in passengers of up to 10 per cent were recorded.

The oil crisis also severely affected developing countries. In 1973 the developing countries earned sufficient foreign exchange from tourism to pay for all imported oil. In 1974 the revenue gained from tourism ($7 billion) was only enough to pay half the oil bill (Waters 1974: 24). This increased their balance of payments deficits and pushed them further behind the economic standards of developed nations.

The continued growth of the airline industry is expected to increase the demand for aviation fuel by between 7 and 8 per cent over the next decade. With world demand increasing, constraints will be imposed on oil availability and competition for the scarce resource will increase. The airlines depend upon oil products and they are unlikely to seek substitutes for oil as an energy source. Given the above conditions, two consequences seem inevitable:

1. Fuel will become increasingly more expensive;
2. Fuel will not be available in the quantities anticipated by many airline planners.

The cumulative results are predictable. As airlines pay more for fuel, they will charge more for tickets, minimize on-board services, and reduce or abolish excursion rates. This, in turn, may eliminate many potential tourists, and reduce the frequency and distance of trips taken by others.

A third consideration is that inflation is increasing the capital requirements and running costs of the tourist sector. The high costs of raw materials, land, labour and energy, increases in interest rates on loans and shorter time-periods imposed for loan repayments have pushed up the development costs of hotel building, airport construction and other infrastructural requirements. These will be offset by transferring them to the consumer and will result in an increased price of hotel rooms, more expensive transportation to and within destination areas, and an escalation in the cost of their tourist goods and services. Price changes will influence the choice of tourist destinations and spending behaviour within them. Tourists are sensitive to price, as is evident in the success of package tours and the emergence of moderately priced hotel chains such as the Super 8 and Budget Inns. Increased prices may contribute to a decline in more expensive forms of travel and accommodation.

A fourth constraint is the rising demand to include environmental, social and cultural issues in tourist development decisions. As tourist numbers increase, demands for more aircraft, airports, hotels, freeways and other services place greater stress on both natural and social environments. In some countries it is now legislated that comprehensive environmental impact assessments must be undertaken for all major development projects. Tourist developments will come increasingly under the compass of such legislation. The result will be delay and increased costs of development. Environmental issues will also become important in existing destinations where crowding has become acute (Edgell 1977: 39). Restrictions have already been introduced which limit the number and behaviour of tourists in fragile natural environments and in crowded historic and cultural sites.

It seems that tourism is likely to become more costly. When, where, and what the effects of the increases will be, are important questions. Will seats on aircraft and hotel rooms become so expensive that only the wealthy can afford to travel? Will many tourist developments currently enjoying record levels of business lose patronage and become so unprofitable that they are forced to lay off workers and close? Will global economic conditions deteriorate to the extent that foreign trade deficits reach their highest levels, prompting governments to impose restrictions on the travel of their citizens? Will petroleum shortages and price increases induce governments to curb weekend fuel sales for motor cars and curtail opportunities for domestic tourism? Numerous questions arise from the above discussion and the tourist industry must carefully consider their ramifications if it is to be in a position to successfully meet the demands which will be placed on it.

Summary

The economic impacts of tourism have been well documented in the literature but considerably more is known about the economic benefits of tourism than the associated costs. The research emphasis on the positive economic impacts of tourism has contributed to the widespread optimism among policy makers concerning the potential of tourism in stimulating economic development. This attitude reflects the generally favourable impacts on the balance of payments, employment and income and, to a lesser extent, entrepreneurial activity. These economic benefits have been accompanied by a variety of costs which, until recently, have been largely ignored in the literature. High leakages from the economies of developing countries, high inflation and land speculation in destinations, low returns on investment because of seasonal fluctuations in demand, and overdependence have been lodged as major criticisms of the tourist industry. Future examinations of the economic impacts of tourism should adopt a more balanced approach which assesses both the benefits and the costs of tourism development.

4 physical impacts

In the past decade the term 'environment' has emerged as a catch phrase to which many diverse and often emotional statements have been assigned. With the passage of environmental legislation there have been renewed attempts to define the term systematically. This has often resulted in the adoption of extremely broad definitions of environment. Environment now includes not only land, air, water, flora and fauna, but may also encompass 'people, their creations and the social, economic and cultural conditions that affect their lives' (Lerner 1977: 2). This chapter adopts a more narrow focus. This part of the book is devoted to physical impacts, for economic and socio-cultural impacts are considered in other chapters. It will involve an examination of the effects of tourism on elements of the natural environment and on the man-made or built environment. The creation of environments for tourism, as in the development of resorts, is also of concern. The environment, be it predominantly natural or largely man-made, is one of the most basic of resources for tourism. However, the growth of tourism leads, inevitably, to modifications of the environment.

Many authors have stressed that man should treat his environment with greater respect than has usually been the case in the past (Carson 1962; Ehrlich 1970; Nicholson 1970). The limited number of studies of the environmental impacts of tourism indicate that this conclusion is applicable here. However, evidence of the environmental effects of tourism is both sparse and scattered. Little systematic work has been undertaken and this is surprising for, as Cohen (1978: 217) has pointed out, this has occurred: 'despite the current preoccupation of scientists, politicians and the general public with the impact of man on the environment and despite the growing interest in tourism as a geographical, social and economic phenomenon'.

The Organisation for Economic Co-operation and Development (1980) has recently published a concise review of the impact of tourism on the environment but their definition of environment is extremely broad and only parts of their report discuss the physical impacts with which we are now concerned.

The state of research on the impacts of recreation on the natural environment has been examined by Wall and Wright (1977). Their monograph did not make explicit reference to the activities of tourists but it can be assumed that tourists were responsible, at least in part, for many of the environmental impacts which were documented. It is unnecessary to reiterate what has been said in this earlier publication and not all of the effects mentioned by Wall and Wright will be considered here. In fact, in many cases, no specifically touristic study could be found which considered the topics discussed by Wall and Wright.

The literature on the environmental impacts of tourism has similar conceptual and methodological difficulties to those concerned with recreation as a whole.

1. Research on the impacts of tourism is topically uneven, and is particularly sparse on soils and on air and water quality;
2. Most studies refer to the effects of tourism on one particular environmental component. There has been little attempt to integrate the effects on a number of components to provide an assessment of the impacts of tourism on the environment as a whole. The components of the natural environment are closely interrelated and highly interdependent. As the activities of tourism are likely to affect more than one environmental component at a time, it is imperative that studies examine the environment as a whole and not individual components in isolation. In practice this is extremely difficult to do and, in order to facilitate presentation, each component will be discussed separately here;
3. Research has had varied regional emphases with most research being conducted in Britain and North America. Studies in Africa have emphasized wildlife whereas those on the Mediterranean littoral have concentrated upon water quality. There has also been an emphasis on particular environmental effects on special ecosystems, such as coasts, islands and mountains. It is extremely difficult to compare the findings of such studies for each location differs ecologically and in the intensity and duration of tourist activity;
4. Much of the research on the environmental impacts of tourism is of recent vintage and has been limited to 'after the fact' analysis. As such, it has the methodological problems which face this type of investigation (Wall and Wright 1977: 4). Such problems include:
   (a) the difficulty of distinguishing between changes induced by tourism and those induced by other activities;
   (b) the lack of information concerning conditions prior to the advent of tourism and, hence, the lack of a baseline against which change can be measured;
   (c) the paucity of information on the numbers, types and tolerance levels of different species of flora and fauna. LaPage (1974: 237) concluded that this makes it impossible to reconstruct the environment in relation to various levels of use, both past and present;
   (d) the concentration of researchers upon particular primary resources, such as beaches and mountains, which are ecologically sensitive.

The above problems should be borne in mind when the results of research are examined.

Tourism – environment relationships

The history of tourism clearly indicates that the environment of places has contributed to the birth and progress of tourism. Scenic sites, amenable climates and unique landscape features have had an important influence upon the patronage of specific localities, regions or countries. The environment of the host region exerts an attraction so that the tourist comes. It offers things that the tourist is looking for and needs (Burmeister 1977: 21). The effects of environmental factors can be observed at all points of the tourism process. Tourist behaviour patterns, although not necessarily motivated solely by environmental conditions, are still clearly influenced by them through choice of destination and length of stay. In other words, climatic features and newly discovered or developed 'natural' attractions may influence tourists' loyalty to, substitution of and behaviour in tourist destinations. Environmental conditions place constraints on types of development and destinations lacking appropriate climatic, geological, floral or faunal conditions are seldom selected for tourist development. Some complementary natural characteristics are highly desirable even if a place has a rich history or is unique in archaeology or culture. Built amenities and infrastructure are usually insufficient, by themselves, for the development of tourism.

Concern for the environmental impacts of tourism has not been restricted to recent writing, although the majority of the literature was written during the past decade. As early as 1961, both Beed and Clement (p. 167) expressed worries that the tourist invasion of Tahiti could induce ecological imbalance within the island's ecosystem if it was not carefully and resolutely regulated. Their concern was expressed at a time when Tahiti enticed a mere 10,000 tourists annually. Now, with an excess of 15,000 tourist arrivals each year, the same fears are being voiced, but little action has been taken to manage and control the situation.

Most research has been reactionary in nature, being a response to immediate threats to the environment. Such threats have been a stimulus to research but they have also resulted in a concentration on special environments, such as small islands, coral reefs and other delicate ecologies. Furthermore, studies have been largely of effects which are highly pronounced and in stages in which management strategies may offer only a slender prospect of returning the ecosystem to its original or near original condition. This research can be viewed as an outgrowth of the environmental movement of the 1960s when widespread concern was expressed over the deleterious, polluting effects of man's activities. Most studies were undertaken after, rather than before, damage occurred. As a result, few studies attempted to elucidate the processes of environmental change or relate these to aspects of the agent of change which, in this case, is tourist development. Ecological investigations are being carried out in many countries but few studies incorporate the role of the tourist as a means by which change is produced. Similarly, tourist developers have often failed to embody ecological principles into development plans and policies promoting tourist activity. A marrying of these research areas is required if a harmonious relationship is to be fostered between tourism and the environment.

To some authors, tourism provides an incentive for the restoration of ancient monuments and archaeological treasures, and for the conservation of natural resources, as well as the economic means by which they can be achieved (Waters 1966; Agarwal and Nangia 1974; Dower 1974). To others, however, tourism means people, congestion, noise and litter. It means the disruption of animal life cycles, the extinction of fragile plants, and the pouring of human waste into rivers and upon beaches (Goldsmith 1974; Crittendon 1975; Mountfort 1975; Tangi 1977). These perspectives represent divergent themes in the literature on environmental effects.

Budowski (1976: 27) suggested that three different relationships can exist between those promoting tourism and those advocating environmental conservation. These relationships are especially important because tourism is highly dependent upon values derived from nature. The relationships are as follows:

1. Tourism and environmental conservation can exist in a situation in which both camps promote their respective positions, remain in isolation and establish little contact with each other. This situation is unlikely to remain for a long period because of the substantial changes in the environment which are apt to occur with the growth of mass tourism. This stage, therefore, is usually succeeded by either symbiotic or conflicting relationships;
2. Tourism and conservation may enjoy a mutually supportive or symbiotic relationship where they are organized in such a way that each benefits from the other. From the perspective of the conservationist, environmental features and conditions are left as close as possible to their original state but, at the same time, they provide benefits to the tourists who view and experience them. There are few places where this has been achieved;
3. Tourism and conservation can be in conflict, particularly when tourism induces detrimental effects to the environment. Most documented relationships between tourism and environment fall into this category. On some occasions, effects of tourism have stimulated conservatory measures in order to protect fragile ecosystems. More commonly, the damage has already reached irreparable proportions.

This chapter concentrates on the latter two of Budowski's three relationships and examines the two major viewpoints held by authors writing on environmental impacts: that tourism is beneficial to the environment and, to the contrary, that they are in conflict.

Tourism and the environment: a symbiotic relationship

Conservation grew from a number of independent roots of which Gunn (1978a: 3) has listed four:

1. There was a social concern to which the park movement owes its beginnings. The growth of industry and commerce and their associated ills stimulated a demand for parks and open space. The provision of public land was seen as an antidote to the immoral values of urban society and as an escape from the routine of work and urban living;

2. There was an emphasis on the efficiency of resource use, particularly of non-renewable resources. Early expressions stressed maximum utilization but with a minimum of environmental degradation;
3. Conservation also incorporated a notion of aesthetic enhancement. This is particularly significant in a recreational context. One major tourist activity is sightseeing which depends heavily on the qualities of the natural environment;
4. More recently conservation has gained a scientific and ecological emphasis in which the maintenance of a balance between man and environment is of prime importance. The outcome of this perspective is the application of stringent controls to protect and preserve the natural environment from unsystematic and unplanned human manipulation.

These roots of concern have contributed to contemporary definitions of conservation. It is now generally agreed that wise and controlled use of the earth's resources is desirable, although tensions exist between proponents of use and advocates of preservation. One result of the environmental movement is that increasingly effective action is being taken, or is envisaged, to control man's impact upon the environment.

Conservation and the preservation of natural areas, archaeological sites and historic monuments have emerged as important spill-over benefits of tourism. In turn, the protection of these prime tourist resources enhances and perpetuates tourism by maintaining its very foundation. The tourist industry has as much interest in maintaining a quality environment as organizations specifically dedicated to that cause.

Wildlife and forest reserves have been established and large tracts of scenic land have been preserved partially because of their ability to attract tourists. Conversely, local tourist offices, tour companies and hotels complement their own facilities by organizing guided hikes and camping trips along authorized trails. In this way tourists, who on their own may have little interest or lack the means to engage in such activities, are able to enjoy viewing wildlife, vegetation and scenery.

Tourism can also be credited with extending environmental appreciation. The tourist industry has discovered, made known and rendered accessible, specific regions and aspects of nature. In Switzerland, for example, mountain vacations in winter were relatively unknown until after the development of mountain railways and sports resorts. Also, the use of mineral springs, of which Switzerland has many, has increased over the past few decades of rapid tourist growth. Similar examples can be found in Canada with 'ecotours' into the Canadian tundra, an area not visited previously by people other than scientists and hunters.

In the absence of an attractive environment, there would be little tourism. Ranging from the basic attractions of sun, sea and sand to the undoubted appeal of historic sites and structures, the environment is the foundation of the tourist industry. Tourism, if it is to be successful and sustained, actually requires the protection of the scenic and historical heritage of destination areas. The protection of such prime attractions has come to be viewed as an investment as the economic potential of tourism has become more widely recognized. This is also a compelling reason for planning and development agencies to concern themselves with environmental issues. Cohen (1978: 219) concluded that such actions are, more often than not, politically defensible since conservation and preservation measures can be presented to the public as being economically necessary and not simply as a means of satisfying tourist demands.

Tourism can assist conservation more directly than by merely promoting its initiation and continuation. Tourism provides both the incentive for conservation and the economic means by which such measures can be carried out (Waters 1966: 116). Part of a region's income from tourism can be invested in the maintenance of scenic areas and historic sites. Beck and Bryan (1971: xxi) stated in their report on tourism in Britain that: 'Many historic houses, villages, old churches and so on could not be kept in a proper state of repair without tourist money.... And given a reasonable influx of tourist money it is usually possible... to keep the tourist nuisance at an acceptable level.'

It is difficult to determine the degree to which tourism has been the sole incentive for the adoption and expansion of conservation measures. Some forms of conservation existed before the advent of mass tourism. However, there is evidence that tourism has been an important stimulus to conservation in developing areas, especially with the support of UNESCO, the World Bank and the World Tourism Organisation. In the light of Europe's magnificent heritage of cities, cathedrals, castles, gardens, churches and monuments, it should not be surprising that most evidence of tourism stimulating the conservation of such sites is from that part of the world.

Conservation within tourist settings has taken four forms. Firstly, tourism has stimulated the rehabilitation of existing historic sites, buildings and monuments. Alexander (1953: 323), in his economic study of Cape Cod, remarked that the quaint villages were the area's major tourist asset and that with a general face-lifting and refurbishing of lighthouses, harbours, fishing piers and old houses, the area was given additional appeal. Williamsburg, the eighteenth-century capital of the British colony of Virginia, is an example of a city which was almost in ruins but has been rejuvenated by the processes of preservation and restoration. Williamsburg is also instructive in its construction of replicas of the past where the original buildings were beyond repair. Newcomb (1979) and Konrad (in press) have provided detailed discussions of the legacy of the past as a recreational resource.

Secondly, tourism has stimulated the transformation of old buildings and sites into completely new tourist facilities. Old cellars and warehouses in many tourist resorts have been converted into quaint bars, discos and restaurants, on many occasions preserving the original structural characteristics. This form of conservation has been favoured in Britain. Middleton (1971: 37) suggested that not only does it bring new life and vitality to declining towns and villages of character and charm – it also relieves existing tourist circuits of increased congestion from the growing number of tourist arrivals. If tourism is to be used in this way co-operation is required between conservation groups and publicity agencies to ensure that the peculiar qualities of the historic towns have been identified and to decide to which sections of the tourist market they can be appropriately directed.

Thirdly, tourism has provided an impetus for the conservation of natural resources. The establishment of national parks in Africa constitutes a good example of the ability of tourism to stimulate conservation. Before Tanganyika, now the Republic of Tanzania, received independence in 1961, it was predicted that African wildlife would be destroyed once European control was relaxed. Not only have such fears been unfounded, national parks in Tanzania and other parts of East Africa have increased greatly in number and strength (Owen 1969: 156). Increasing recognition by East African governments that tourism may contribute to foreign exchange earnings has stimulated a corresponding realization of the need for the conservation of natural resources which appeal to tourists, particularly wildlife. The numerous parks which have been set up in the last twenty years in savannah Africa have been established in spite of the presence of economic, political and social pressures, and they confirm the place given to tourism as a means of economic development. More than 80,000 square miles (207,200 square kilometres) have been set aside as national parks in East and South Africa, harbouring one of the world's last and greatest wildlife populations. Serengeti alone is 6,000 square miles in area (15,540 square kilometres) and is the home of more than one million wild animals, including at least thirty species of grazing animals and twelve different predators (Pollock 1971: 146). Lake Manyana National Park, lying at the foot of the escarpment of the Great Rift Valley, is a well-known haven for wildebeest, water buck, giraffes, zebras, lions, leopards, and flocks of flamingos and pelicans (Netboy 1975: 27). Tourism has fostered the protection of these diverse species as they serve as major attractions for both domestic and international tourists.

The establishment of recent parks has required additional justification to that of wildlife preservation. Tourism development, through the creation of parks, generates other benefits as secondary effects of conservation. These include net gains in foreign exchange, the creation of employment within the park and in ancillary tourist services, the fostering of demand for locally made products, and the hastening of infrastructural construction. Such effects have often been employed as justifications for the expansion of African national park systems. Moreover, it has been argued that the economic benefits derived from the use of such areas for tourism far exceeds those which would be yielded from agriculture. Netboy (1975: 27) stated that: 'Africa has an enormous comparative advantage over the rest of the world in producing wildlife and can do it on land unsuited to almost anything else.'

In somewhat similar fashion the island of Dominica, a place with many pressing economic and social problems, has set aside parks. Thorsell (1973: 19), recapping the Dominican experience, supported the ideal of parks serving the dual purposes of preservation and the enhancement of human enjoyment. He stated that 'Parks must be seen to contribute to the total economic development process by providing a flow of multiple benefits. . . .' Within the broad scheme of wildland management, of which tourism is an integral part, he postulated that parks can play their most effective role as environmental catalysts. By this he meant that, through education, parks may promote environmental sensitivity in other economic sectors. However, the recent writings of Myers (1973b, 1975), Swift (1972) and Mountfort (1975) have seriously questioned the role of tourism as a mainstay of wildland parks. Their arguments and conclusions are presented in a subsequent section on wildlife.

Fourthly, tourism has been responsible for the introduction of administrative and planning controls which have been adopted in order to maintain the quality of the environment and to ensure the provision of satisfying experiences for visiting tourists. Measures have varied from site to site and from country to country depending upon the resilience of the protected area or attraction, the nature and intensity of tourist use, and the political structure of the bodies initiating and administering the controls. Unfortunately, many such measures have emerged as a result of the physical deterioration of sites and attractions from uncontrolled and excessive use. For example, in Britain, Hadrian's Roman Wall is crumbling and Westminster Abbey is wearing out under the pressures of too many tourists (Jensen 1979). Similar problems are being faced on the Great Wall of China, in Yosemite National Park, and at the Parthenon on the Acropolis of Athens.

The controls which are adopted vary with the seriousness of the problems which are to be faced. In many historic areas in Europe, traffic has been restricted adjacent to major attractions, access to popular venues has been controlled, advertising in public places has been curtailed, and incompatible development has been prevented. Modifications of such measures have been applied in the national parks of Africa and Australia where vehicular traffic is restricted to ecologically desirable routes, development within parks is prohibited, and activities which create detrimental effects are curtailed. In extreme cases, where tourist numbers are large and visitation pressures are intense, more radical conservation measures have been employed. The tourist attractions of Stonehenge in England and the Parthenon of Athens have recently been designated 'off limits' to tourists, and access to the interior of these sites has been eliminated. In 1974 a General Plan of Controlled Growth in Hawaii mandated that tourist developers are to complete detailed surveys if there are indications that their projects are likely to affect important historical and cultural areas. Most of the more moderate measures which have been employed to control the relentless pressures of tourists have met with only minimal success. As tourist numbers rise and visitation to ecologically sensitive sites increases, the demands for more extreme conservation measures will intensify.

In summary, on some occasions tourism has been a force for the preservation and rejuvenation of areas by stimulating the conservation of historic and archaelogical sites, and unique or quaint man-made landscape features. It has also been an incentive for the conservation of natural resources which, because of their uniqueness or educational or ecological values, are appreciated as tourist attractions. This type of conservation leads to the growth of tourism in these destinations. In time, subsequent conservation measures may ultimately be required to maintain the quality of the environment and protect it from the increasing number of tourists. The extreme actions taken in Greece and in Britain, and the new regulations in Hawaii, are indicative of a growing feeling of uneasiness that the benefits of tourism may not necessarily compensate for the costs of conservation. Some authors argue that through its stimulation of conservation, tourism has been instrumental in awakening an appreciation for nature and history (Huxley 1961; Dower 1974). Turner and Ash (1975), Haines (1976), Jensen (1979) and others have opposed this viewpoint and have indicated that the conversion of historical monuments, archaeological ruins and natural areas to tourist attractions robs them of their magic but attracts more tourists who place undue pressures on these attractions through their physical presence. This, in turn, creates a demand for further remedial action. Nevertheless, there is little evidence to indicate the widespread existence of a symbiotic relationship between tourism and the environment. The majority of the literature examines relationships between tourism and environment in conflict. This relationship is discussed below.

Tourism and environment in conflict

The general texts on environmental quality are full of references to air pollution from car exhausts, the destruction of coastlines, the pollution of rivers from human wastes and detergents, and traffic congestion. These texts rarely mention that tourism is partially responsible for such effects. The negative environmental ramifications of tourism have received attention in the literature only recently. According to Charlier (1977: 141), a recent book by Haulot considered a wide range of effects of tourism on the environment and, in doing so, established a precedent. This part of the chapter will discuss the impacts of tourism on particular environmental components. Aspects of the natural environment will be considered first and this will be followed by an examination of the effects of tourism on a number of different ecosystems. The effects of tourism on the man-made environment are presented in the latter portion of the chapter.

Tourism and vegetation

Vegetation is one of the major attractions of many destination areas. The sequoia redwoods of California, the kauri forests of northern New Zealand and the Black Hills spruce of South Dakota are examples of vegetation which have allure for tourists. In spite of vegetation being a primary tourist resource, little is known of the effects of tourism on vegetation. However, there is a large body of literature on the effects of recreation on vegetation and much of this is also applicable to tourism.

Most of the literature has concerned relationships between tourist behaviour and the magnitude of physical damage caused to vegetation. A variety of tourist activities impact upon vegetation. They include the following activities and effects:

1. The collection of flowers, plants and fungi can result in changes in species composition;
2. Careless use of fire in parks has caused major conflagrations in the forested regions of California and Australia;
3. Deliberate chopping of trees for tent poles and firewood. This has removed many younger trees from forests which alters the age structure of the plant community, leaving fewer trees to mature and provide shelter for the site (Wall and Wright 1977: 26);
4. Excessive dumping of garbage, which is not only unsightly but also changes the nutrient status of soils and can be ecologically damaging by blocking out air and light;
5. Pedestrian and vehicular traffic impact directly on vegetation. The impact becomes a problem when the intensity of use is excessive and this depends upon the vulnerability and capacity of the ecosystem. For example, in flat areas with compact soil effects may be minimal when compared with those on dunes. Mather, Ritchie and Crofts (1973: 865), working in British coastal areas, stated that in addition to the destruction of dune vegetation, linear blow-outs developed along the weakened zone. Such blow-outs can expand rapidly and may extend inland for large distances. This reduces the area of land for grazing and recreational uses. Westhoff (1967: 222) claimed that trampling had destructive effects on the sequoia redwoods of California. Constant trampling around the base of the trees damaged roots and decreased the quantity of fungi;
6. Camping has similar effects to trampling. The construction of campsites involves the removal of vegetation and, as Noake (1967: 226) pointed out, the damage also extends into the surrounding area with the development of trails and picnic sites.

The above activities have other impacts in addition to those resulting from mechanical damage. As Wall and Wright (1977) have noted, there may be changes in percentage cover, species diversity, growth rates and age structures, and habitat diversity. No studies were located which ascribed these effects specifically to tourism. If it is assumed that the documented effects of recreation on vegetation are also applicable to tourism, then a number of conclusions can be drawn:

1. The greatest damage to vegetation cover occurs with the initial use of an area. The most fragile species disappear with initial use and recovery is dominated by more resilient species;
2. There is a decline in the diversity of species with continued use. Only the most tolerant species survive and other tolerant varieties may be introduced;
3. Some vegetation cover in grassland ecosystems registers little deterioration because of its high proportion of resilient species. Hence the effects of tourism will vary greatly from ecosystem to ecosystem;
4. The reproduction rates of vegetation are greatly reduced in trampled areas;

5. There is a strong relationship between soils and vegetation. Soil compaction will influence plant growth and the age structure of vegetation.

Although Watson (1967) and Bayfield (1971, 1974) have undertaken some work on ski areas, there is little evidence in the literature of the effects on vegetation of large tourist developments such as hotel complexes and parks. It is imperative that vegetation, as a major tourist attraction, be protected from the adverse impacts of tourism. Research is required to extend the work which has already been done on recreation. Research on recreation has concentrated upon specific, small sites. Tourism research should also examine effects upon large tracts of vegetation.

Tourism and water quality

No studies have been located which examine the effects of tourism on water quality. This situation has arisen in spite of the existence of a small but growing body of literature on recreation and water quality, and the growing concern regarding the pollution of recreational lakes, rivers and coastlines. A number of studies mention that water pollution is becoming a problem in coastal resorts, but detailed analyses have not been pursued beyond these general statements (Clare 1971; Tangi 1977).

The waters of the Mediterranean are an asset to the tourist industries of Italy, Greece, Yugoslavia Spain, Tunisia and France. Unfortunately, the Mediterranean is also a repository for municipal and industrial waste, for oil spills from tankers, and for pesticides brought down by the rivers which flow into the sea. Tourist developments along the coasts of the Mediterranean are also responsible for large quantities of waste materials. In this situation, tourism threatens to undermine the resource which has been primarily responsible for its existence. Tangi (1977: 336) noted that diseases such as cholera, typhoid, viral hepatitis and dysentery can be transmitted through contaminated seafood. There is a very real threat that this could occur around the Mediterranean. Water pollution is now reaching crisis proportions in a number of the older tourist resorts.

Given the seriousness of the situation, it is surprising that more research has not been undertaken to establish quantitative cause–effect relationships between tourist activity and water quality. For activities such as swimming, fishing and many types of boating, high quality water is essential. The introduction of pollutants into water bodies is both environmentally damaging and economically disastrous for water-based tourist resorts. Again, assuming that the recreational pursuits discussed by Wall and Wright (1977) will also be pursued by tourists with similar effects, their conclusions will be applicable to tourist activities as well. The major effects listed by them were as follows:

1. Pathogens are introduced into the aquatic environment in raw or inadequately treated sewage. The release of sewage onto beaches, lakes and rivers is often mentioned in the literature on environmental impact. It is a potential health hazard to tourists using those resources;

2. The addition of nutrients to the water will hasten the process of eutrophication. Excessive weed growth may result with implications for levels of dissolved oxygen which, in turn, govern fish numbers, species composition and growth rates;
3. Fluctuations in the supply and distribution of oxygen have implications for aquatic plant and animal life which are well understood. The presence of increasing quantities of oil from recreational vehicles, and oil spills from tankers and cruise ships in resort harbours, have had detrimental effects on oxygen supply and have reduced the satisfaction of tourists swimming in those waters;
4. Petrol from recreational vehicles can increase levels of toxicity in lakes and rivers and will result in detrimental effects to aquatic plants and wildlife. The lead compounds which are used in petrol accumulate in the bottom sediments and become toxic to some life forms. The effects of detergents and of trace elements from the decomposition of tins and bottles are also thought to be harmful but their consequences have yet to be fully researched.

Most of the research which has been conducted on water quality has examined the effects of urban and industrial wastes. Although the principles may be similar, tourist activities inject different types of pollutants, in different quantities, into water bodies which are of high quality. The repercussions are also likely to differ. The application of quantitative analyses to determine the nature and magnitude of such repercussions is urgently needed.

Tourism and air quality

There is a dearth of material on the impacts on air quality of recreation in general, and tourism in particular. As tourism involves travelling, normally by motorcar, ship, train, bus or aeroplane, the contribution of each to air pollution is of relevance to the theme of this section. For example, the concentration of pollutants from motorcars within cities is attributable partially to recreational driving, tour buses and taxis. In resorts, the pollutant contribution from tourist vehicles is likely to be significant. One can only speculate on the importance as figures are not readily available. Viewpoints on the subject vary. Adverse impacts on vegetation have been attributed to motorcar exhausts in congested valleys, such as Yosemite. In contrast, Soo Ann (1973: 207), in a paper to the Pacific Area Travel Association supporting the theme that 'Tourism Builds a Better Environment', claimed that tourism has a smokeless economic base. Tourism does not involve the movement of large quantities of raw materials and manufactured goods, or the transformation of one into the other with its attendant pollution. Rather, it involves the movement of people. As a result, it is claimed that it does not pollute the environment as much as most industries. Houghton-Evans and Miles (1970: 425) concluded that the impact of motorcars on the air quality of rural areas is probably insignificant except in congested spots. There appear to be no studies of the problem in resort areas where motorcars concentrations are high. Considering that traffic congestion is a major problem in many resorts, this is somewhat surprising. Although the contribution of tourism to air pollution may be less than for some other forms of human activity, this is no reason to disregard it entirely.

With the exception of intra-European tourism, the airlines are a major mode of travel for international tourists. The aircraft is of great significance to the tourist industry but it contributes little to the destruction of air quality. Air transport in the United States is only a minor source of pollution, producing only 1 per cent of all air contaminants. Studies undertaken at London's Heathrow Airport, and at Tokyo and Los Angeles, showed that carbon monoxide levels were less than one third of those recorded in the downtown areas of those cities. Shaw (1973: 227) concluded from the Heathrow study that: 'far from adding significantly to pollution levels in adjacent areas, the airport dilutes the pollution from surrounding districts and reduces concentrations of pollutants below the levels which might be expected if an average urban development took place'.

In spite of increases in the volume of traffic, the new designs of jet engine combustion chambers have reduced pollutant emissions. The gradual replacement of propeller planes, and improvements to short-haul B727s and DC9s, have reduced smoke emissions and invisible gaseous pollutants. The aircraft has been insignificant when compared with the motorcar as an agent of change in the chemical quality of the air. This conclusion does not hold true for the contribution of aircraft to noise pollution. This is a major environmental problem of aircraft and satisfactory remedies have yet to be devised. The reduction of aircraft noise by 10 decibels involves a 90 per cent cut in sound output. The technology is not yet available to allow this to occur without also lowering the safety margins of the aircraft's engine thrust. Indirect attempts to control noise by altering flight procedures and reducing the number of night flights have allowed those living near airports to experience a decrease in annoying noise and reduced some of the psychological stress associated with high noise levels.

Tourism and wildlife

The hunting of animals and, more recently, the viewing and photographing of wildlife are important tourist activities. The rapid increase in patronage of East African national parks (30 per cent per annum over the past fifteen years), the emergence of substitute safari parks in North America, Europe and Australia, and the continued popularity of zoos bear witness to the fascination with animals. The quality of observation is of great importance for tourist satisfaction and, according to Reid (1967: 77), a high calibre of experience is most likely to be achieved when a wild animal is viewed, undisturbed, in its natural environment. Viewing and experiencing the splendour of African wildlife in a natural setting exceeds the artificial spectacle experienced in city zoos. The almost universal desire to view wildlife in as natural a setting as possible concentrates large numbers of tourists into a limited number of areas endowed with rich wildlife habitats. It is also under such conditions that the most significant environmental impacts occur.

Measurement of the impacts of tourist activities on wildlife have been plagued by the lack of accurate population counts prior to the advent of tourist activity, and the lack of distinction in counts between out-migration and mortality (Wall and Wright 1977: 41). There are few longitudinal studies of wildlife populations, particularly investigations which note effects upon population composition and consider relationships between cause (tourist activity) and effects (changes in species numbers and composition).

Examinations of the impacts of recreation and tourism on wildlife have emphasized animal species in relatively natural environments. Tourism studies of wildlife differ from those in other aspects of recreation in three major ways:

1. The documentation of impacts of tourism on wildlife has concentrated upon larger mammals and birds. Little attention has been directed to the effects of tourism on small mammals, reptiles or insects;
2. Studies have tended to be at a macro rather than a micro level: for example, in national parks and nature reserves. This is understandable as wildlife preservation areas are designated on this scale and serve as primary tourist resources;
3. Studies of the impacts of tourism on wildlife lack a firm ecological base from which effects can be determined and predictions can be made. Most studies are descriptive and make only fleeting reference to ecological principles of migration, feeding and breeding behaviour, and adaptation to changes in habitat. Studies by Myers (1973a and b, 1975, 1976) and Carbyn (1974) are notable exceptions.

This section is primarily concerned with the effects of tourism on wildlife living in natural surroundings. Most examples are taken from studies performed in East African national parks and game reserves where relationships between tourism and wildlife are most evident and where wildlife forms the prime attraction. Tourism has been a major stimulus to the preservation of wildlife, especially in Africa, and this is a highly positive manifestation of tourist development. Nevertheless, recent writings on tourism and wildlife have painted a less rosy picture than this might suggest. Some authors (Burn 1975; Rensberger 1977) have even predicted that if tourism is allowed to continue as the major rationale for the existence of African national parks, they have only about fifteen years of useful life remaining. Rivers (1974: 12) discussed a number of effects which form the basis of such predictions. These effects can be divided into two categories: direct and indirect impacts.

Direct impacts on wildlife

The ability of wildlife to withstand an influx of tourists varies from species to species and from region to region. For example, according to Curwen (1973: 7), wildlife in Botswana could tolerate considerably higher numbers of tourists without incurring disruptive effects. Other authors (Mountfort 1974;

Myers 1976) have noted that carrying capacities have been exceeded in the developed African national parks of Serengeti, Tsavo and Mount Kenya. In these parks the density of tourist vehicles within a 10–15-mile radius of tourist lodges has become so excessive as to disrupt the tranquil environment. This also detracts from the tourist experience. Hutchinson (1975: 107) suggested that: 'Much of the attraction of these parks lies not so much in the presence of animals as in the absence of humans and tourists become quite indignant when they see other tourists viewing animals.' Reporting on a recent safari experience, Gray (1973: 26) noted her frustration in attempting to photograph animals without including other safari vehicles. A recent newspaper article (Kitchener-Waterloo Record 1979) typified the changes in African game parks as follows: 'In the old days, Masai herdsmen knew lions were near when their long horned cattle stopped eating and all gazed in the same direction. Today, tourists to Amboseli National Park in the heart of Masai tribal lands can find the king of the beasts by looking in the centre of a ring of idling minibuses.'

The direct effects of tourist activity on wildlife depend largely upon the intensity of tourist development, the resilience of species to the presence of tourists, and their subsequent adaptability. Some of the major effects which have been mentioned in the literature are summarized below.

The disruption of feeding and breeding For many tourists the observation of a predator stalking and securing its kill is the highlight of a wildlife safari. Gray (1973: 26) dramatically described this as follows: 'that people are travelling to Africa not only to bathe in nature's innocence but also to witness nature's violence. To see predators on the kill is considered to be the great moment of a safari, a new outlet for the blood lust once channelled into hunting'. This lust frequently undermines the privacy of the animals being observed. Gray (1973: 28) noted that on one occasion tourists in a safari van watched a cheetah stealthily approach its prey for twenty minutes. Immediately prior to the critical point in the cheetah's approach, the vans converged and tourists climbed on the roofs to take photographs. They created such a commotion that the prey, an impala, became startled and ran off. Myers (1975: 6) claimed that the close shadowing of lions by tourists leads to many missed kills with the result that many lion cubs are starving. He also cited occasions when young animals became fatally separated from their mothers because of the erratic behaviour of minibus drivers. This behaviour is often encouraged illegally for drivers are offered large sums of money to break park rules and chase animals. It seems that the chasing of animals has increased markedly in recent years and this activity has caused numerous cheetahs to die of heart failure.

Mountfort (1974: 7) noted the disruptive effects of over-enthusiastic and thoughtless photography. The pressures of tourists taking photographs has caused a noticeable decline in the breeding success of many species of coastal birds in the Galapagos Islands. Tourists, insisting on photography at close quarters, have scared birds away from their nests so that the eggs and the young were abandoned or taken by predators. According to Mountfort, visitors' handling of the creatures has also caused an increase in infant mortality.

The development of highways and trails through parks, in areas which were traditionally feeding and breeding areas, has forced wildlife to relocate. Carbyn (1974: 98), in a study of wolf populations in Jasper National Park, found that many of the dens which were in close proximity (0.5 miles or 0.8 kilometers) to the main highways were abandoned.

The killing of wildlife While indiscriminate hunting and fishing will reduce wildlife numbers, animals are also killed accidentally. Kraus (1967: 161), as a result of a study conducted in a German national park, reported heavy losses to game, especially hares, roebuck and red deer, with an increase in vehicular traffic. Animals are also frequently run over in North American parks. Some species benefit from such accidental mortality. Hawks and falcons are often to be seen scavenging on park roadsides.

The disruption of predator–prey relationships Some studies exist which consider the responses of deer and other ungulates to the presence of tourists. They describe the capability of deer to become habituated so that they freely accept the presence of man and his structures (Carbyn 1974: 98). Carbyn also noted that the grazing of elk and bighorn sheep along the roadsides of Jasper National Park stimulates tourists to stop to take photographs at close range. In addition to being a traffic hazard, this also forces wolves to hunt ungulates in the interior of the park, thereby reducing the availability of potential kills. Increased concentration of wolves in smaller areas will also increase the competition among themselves.

Myers (1972, 1975), Gray (1973) and Mountfort (1974) have speculated that the disturbance of wildlife in Africa emanating directly from tourist activities are of a significance comparable to that from poaching. If this is so, then these disturbances pose a very real threat to the wildlife of the region.

Indirect impacts on wildlife

The creation of national parks and game reserves has been conducive to the proliferation of certain species. Recent monitoring of animal populations in African national parks revealed sharp increases in the number of wildebeest, zebras, hartebeest, gazelles and elephants. Some authors (Myers 1972, 1975; Rensberger 1977) predict that there will be overpopulation in the near future. Natural control mechanisms may take one of two forms:

1. *Increased competition for food.* This will stimulate fighting and will also have adverse, often fatal, effects on the weak and the young.
2. *Mass out-migration to alternative environments.* This form of adjustment has not been observed in Africa for seventy years but some authors fear that such an event is imminent. This adjustment should not be confused with the common phenomenon of seasonal migration.

Increases in one species may also affect the numbers of others. The elephant is a good example. Because of the expanding human population in East Africa, game parks have become the elephant's only safe refuge. The parks are too small to support the increase in elephant numbers (Hutchinson 1975: 104). African elephants are not conservationists. In excessive numbers, through the stripping and uprooting of trees, they destroy the environment on which they feed. This has resulted in the gradual conversion of woodland to grassland. A study undertaken in Tanzania's Great Ruaha National Park led to the conclusion that elephants reduced the tree coverage by approximately 10 per cent annually. This, in turn, reduced the food supply for other species, particularly for the browsing species such as the giraffe and the black rhinoceros.

The extension of tourism also results in changes in wildlife habitats. Litter around camping areas and garbage dumps in parks have attracted fauna, such as bears, birds and rodents, to these areas. By littering, tourists have not only altered habitats, they have also disturbed traditional feeding patterns. Both grizzly and brown bears have been attracted to areas frequented by tourists so that the frequency of human interactions with bears has increased. Injury to tourists by bears has been noted by Moment (1969) and Martinka (1974). Martinka (1974: 15) noted a decrease in injuries to tourists and damage to equipment in Glacier National Park with the implementation of management plans which encouraged bears to return to their normal feeding habits. Timely removal and adequate disposal of garbage, temporary hiking and camping restrictions, increased patrols, and the relocation of troublesome bears, are among the measures which have been adopted. Similar controls are also proving effective in Yosemite National Park (Darling and Eichhorn 1967: 100).

Tourist souvenirs are not restricted to manufactured trinkets. The capture and killing of animals for trading purposes has increased with the growing demands for wildlife souvenirs. These may take the form of furs, skins, ivory ornaments, horns and tails. The curio trade in East Africa is booming and is visible evidence of the destruction of wildlife to satisfy the whims of tourists (Rensberger 1977: 138). The number of curio shops has increased rapidly: 'There are scores of them stuffed with elephant tusks, zebra hides, mounted antelope heads, lion-claw necklaces, monkey-skin throw rugs, ostrich-foot lamps, gazelle-hoof key rings and assorted other animal products eagerly snapped up by wealthy tourists'. Crocodile skins and stuffed baby crocodiles are sold as souvenirs and this has been responsible for the eradication of the species from many African rivers and lakes, and has decreased numbers drastically in the Caribbean and in South America (Mountfort 1974: 7; Pollock 1974: 146). The growing curio trade has occurred in spite of the enactment of legislation to prevent the hunting of wild animals. The question arises, given the existence of hunting controls, as to where the curio shops acquire their supplies of wildlife souvenirs. One legitimate source of a few animals is the government game control programmes. Such programmes permit a limited number of elephants, lions and other animals to be shot if they endanger local inhabitants or their crops. They also collect animals within the parks which have died through natural causes. It is unlikely that these sources could supply the quantities of animals seen in curio shops. Moreover, it is seldom necessary to control such species as zebra, antelope or monkeys. The curio trade is an obvious sign of one of the greatest threats to African wildlife: poaching.

A large proportion of the income from tourism never reaches the African native. As long as their standard of living remains low, no amount of argument or persuasion is likely to stop poaching entirely while the incentive of securing cash for animal trophies is high. Owen (1969: 158) believed that, in spite of the risk of being caught and either fined or imprisoned, poaching is likely to continue. Netboy (1975: 27) stated that in 1973 at least a thousand elephants were being slaughtered every month. This was expected to increase as the trade for ivory grew and its value rose: tusks are currently fetching in excess of US $2,000 on foreign markets. Netboy also noted a case in Kenya where a poaching ring was broken up, thereby halting the annual export of 25,000 zebra skins.

Until relatively recently poaching has been primarily by individuals for meat, skins and some money, and has been limited to a few isolated areas. Hunting techniques have been crude, depending upon poisoned arrows, camouflaged pits and wire snares. As the tourist industry has expanded in Africa and demands for wildlife trophies have increased, more dangerous forms of poaching have evolved. In parts of northern Kenya armed gangs using trucks and sophisticated weapons, including machine guns, have been reported. Some of these organized poaching enterprises are internationally funded.

The threat to wildlife is very real and it is likely to intensify with increases in tourist arrivals, demands for souvenirs and the continuation of low levels of income for local residents. The extent to which this threat will spread throughout Africa is uncertain. It will depend upon the amount of tourist revenue which is redirected into park development and local economies, and the level of priority which is given to anti-poaching measures. In the general absence of effective management plans in wildlife areas where tourists visit and stay, the impacts upon wildlife have been largely negative. Little is known of the effects of particular types of tourist development and activities on the wildlife of surrounding environments, or of the effects on individual species, including bird and fish life, as well as mammals. More research is required on the ecological aspects of wildlife, which should encompass reproductive and feeding habits, species diversity and resilience to habitat change. The application of research on island ecosystems to national parks and game reserves is a potentially fruitful avenue of investigation (McEachern and Towle 1974).

Tourism, wildlife and the needs of local residents

Conflicts between the needs of tourists and the needs of wildlife are only a part of the wildlife problem in East Africa. A complex relationship exists between tourist development, the requirements of wildlife, and the needs of local inhabitants residing in areas adjacent to the boundaries of game parks.

According to Myers (1972), Swift (1972) and Allen (1976), this three-way conflict is responsible for the impending crisis faced by much East African wildlife.

The conditions leading to this conflict merit brief examination. Annual population growth rates of 3.5 per cent are experienced in the African countries harbouring the major wildlife reserves. More than 50 per cent of the population is less than fifiteen years old. The average family size totals between six and seven children. As nutritional levels rise in the segment of females of reproductive age, the number of children per family is expected to continue to rise. The problems of increasing population are compounded by the lack of suitable land for agriculture and overpopulation of existing arable lands: such areas often currently experience population densities of 1,600 per square mile (615 per square kilometre). Together, these two features place great stresses upon existing agriculture. Faced with these pressures, land-hungry Africans have spilled over into the drier, game-rich, savannah grasslands. This is inducing major conflicts between wildlife and the needs of local inhabitants.

Tourism lies in the middle of this conflict. African national parks and wild lands yield a greater return in their natural form than if they were used for cultivation or grazing. In economic terms, the marginal loss of food from declaring Serengeti out of bounds to agriculture is more than compensated for by the gain in utility in having the animals conserved. Tourism has been largely responsible for this. On the other hand, African natives require food for survival and are forced to seek areas on the margins of national parks for cultivation and grazing. The benefits of tourism never reach these people and their attitudes towards preservation are swamped by their attempts to survive.

The areas which are most susceptible to these conflicts are the margins of the major parks. Ideally the zones surrounding the parks should provide a breathing space. Now these buffer zones threaten to become a noose to strangle the life out of the park's wildlife (Myers 1973b: 22). As population continues to increase and more land on the edge of the parks becomes used for agriculture, the threats to wildlife become more apparent. These threats result from:

1. *The loss of habitats*. Instead of finding a buffer zone in the park hinterlands, animals are encountering competition for water and grazing land. In Lake Manyana National Park in Tanzania, elephants have become trapped in the park because of human encroachment on the areas which surround it. Foraging pressures are having serious effects on forests within the park. Ugandan figures indicate that the amount of land available to elephants fell from 70 per cent of the national land area in 1929 to a mere 17 per cent in 1972. Moreover, in some park areas elephant numbers have increased, placing intense pressures on food supplies. Many areas of Africa which were previously uninhabited by humans because of the tsetse fly are now regarded as having a potential for agriculture. Their development would result in a loss of habitat for much wildlife;

2. *Increased pressures on predator–prey relationships.* The restriction of animals to the parks reduces their hunting territories and disrupts their life support systems. Lions and cheetahs, in times when natural prey is in short supply, will turn to cattle and sheep. As a result such predators are threatened by stockmen. Myers (1975, 1976) has written extensively on this problem and offered grim predictions on the plight of the cheetah and lion. Poison has been known to eliminate an entire lion pride in one night. Cheetah numbers are becoming alarmingly low. Africa now has between ten and fifteen thousand cheetahs and this is only half the number it had in 1960.
3. *Restrictions on movement.* Development around park peripheries has prevented mobile species, such as the springbok in South Africa, from migrating. The likelihood of this occurring to other species, such as wildebeest in East Africa, is increasing.

Losses of habitat, and the disruption of predator–prey relationships and migratory behaviour will undoubtedly reduce the total number of wildlife. Unfortunately the species most likely to be affected are those which are already facing problems of survival.

A major task which must be faced by the East African tourist industry is the justification of national parks as a means of meeting the needs of the local community, as well as tourists and nature lovers. A major challenge is to provide land, food and work for a growing population while preserving the wildlife heritage. There are no easy answers. Policies of outright protection of parks have served well enough to date but, given the pressures on the land and wildlife of Africa, such policies may not be in accord with the legitimate needs of the people of the region. Any strategy which threatens the existence of the parks is not acceptable, but if the lives of local residents are in jeopardy because of the inadequate supply of land and food, then policies of strict protection seem equally deplorable.

Tourism and geology

Information on the effects of tourism on geology are scarce. There has been occasional mention of the impact of collectors of minerals, rocks and fossils, and in some areas the stripping of caves of their natural formations has become a serious problem. The extraction of unique rock formations by the professional suppliers of souvenirs, and the wear and tear from large numbers of tourists have both been contributing factors. Vandalism has also been reported in some of the more popular cave sites. The defacement of rock faces by the carving or painting of initials, names and inscriptions leaves unsightly scars but is not usually a major ecological problem. Mountaineers may alter the character of rock faces as a part of their activity. Watson (1967) pointed out that mountaineers are a relatively small group and seem to have little impact on rock faces other than minor abrasions. However, the repeated cutting of footholds can change the character of popular climbs and can detract from the recreational experience of other climbers. Damage has also occurred to the Australian attractions of Ayers Rock (Ovington, Groves, Stevens and Tanton, 1972) and the Great Barrier Reef, and to the coral atolls of the Pacific. One can conclude from the limited number of investigations that impacts on geology are restricted to special environments and are largely concerned with the destruction of unique features. This will not be a major problem in most tourist areas.

Impacts of tourism on ecosystems

An alternative way of examining the impacts of tourism on the environment by a consideration of environmental components, is to focus attention upon distinctive ecosystems. This approach was adopted by Satchell and Marren (1976) in their examination of the impacts of outdoor recreation. The impacts of tourism on a number of distinctive ecosystems will be reviewed here.

Coastlines

In many western countries coastlines have received the full force of recreational pressures. In addition to tourism, coasts are also the sites for other activities such as ports, power generation and refineries. Thus tourism adds to stresses which are already imposed upon fragile coastal resources. The ecological impact of coastal tourism is a complex problem because of its range of beneficial and detrimental effects. In order to make a region more attractive to tourists, measures are sometimes taken to drain swamps and to protect the flora and fauna (Tangi 1977: 338). However, drainage is incompatible with the preservation of many species. Most effects of tourism have been largely negative because of inadequate planning. Detrimental effects include the elimination of some plant and animal habitats, the obliteration of geological features by excavation, water pollution, and a diminution in the aesthetic qualities of scenery. Human pressures inevitably lead to ecological disturbance, disfigurement of the coastline, and a reduction in the attractiveness of the resource. Jackson (1973: 64), Goldsmith (1974: 48) and Crittendon (1975: 10) all claimed, for example, that disposal of wastes in the waters of many resort beaches has reduced the bathing and boating potential of those areas.

The most explicit inventory of the effects of tourism on coastal ecology has been offered by Boote (1967: 131–7). He listed possible impacts according to the geological composition, relief of the coastline and types of tourist activity. For example, the sand and shingle dune type of coastal environment is particularly attractive for such activities as sightseeing (by foot and by vehicle), golf-course development, fixed accommodation, caravan and camping sites, and bathing. The effects arising from sightseeing alone include the disruption of zonations, destruction of habitats, erosion of dunes and interference with the breeding habits of wildlife. Accommodations and camping sites have serious effects, particularly if cluttered in an unplanned fashion. These include problems with litter, sewage disposal, fires and erosion. In salt marshes and mud flats the tipping of rubbish, land reclamation and the building of marinas alter water levels and nutrient concentrations, and disturb flora and fauna.

The above examples are only a limited sample of the diverse environmental effects which are currently occurring in coastal areas fostering tourist developments. The coastlines of Spain provide numerous examples of the deterioration and the destruction of natural ecosystems from unplanned tourist development. Concrete walls along the Costa del Sol are an example of what has been required in the past and what should be avoided. Other eyesores include the construction of skyscrapers, unsightly hotel developments and waste-water processing stations. Rodriguez (1978: 50) believed: 'that large scale tourism has invaded the Mediterranean coasts and is converting this sea into a dead sea where people will soon be unable to bathe unless they want to catch some disease'. Tourism, which has been so helpful in bringing foreign exchange to the countries along the Mediterranean shore, is, in the long run, destroying the sources of its success.

Coastlines are key areas where planning measures and land-use controls must be implemented if that environment is to make a lasting contribution to the tourist industry. Attention must be devoted to environmental health and the purification of waste water, to the prevention of development on unstable areas such as dunes and eroding cliffs, and to the regulation of aesthetically unpleasing architecture.

Oceanic islands

Oceanic islands, like coasts in general, are experiencing pressures from a variety of activities. Because they are often small and isolated they have only limited resilience to environmental disruptions. Clare's (1971) dramatization of the struggle of Australia's Great Barrier Reef against the onslaught of mineral exploitation and tourism is especially instructive for it portrays the susceptibility of coral islands to such activities. Green Island, within the Great Reef, has more than 80,000 visitors each year. Tourist development and the activities of tourists have combined to produce major effects on the island's near-shore ecology. The conditions which have led to this unfortunate situation include the following:

1. Tourists wade out from the beach onto the reef flats at low tide. According to Clare (1971: 45), they are 'walking on living organisms'. A mass of broken coral skeletons on the reef floor are covered with a brown-grey coating of algae. A large proportion of the coral and small fish life around the margins of boat jetties and hotel beaches have been killed;
2. Souvenir shops choked with shells, shell-jewellery and ornaments, and an assortment of dead coral are located on the island close to the point of disembarkation. The removal of many life forms from the reef for the souvenir trade must have a marked impact on the reef ecology. Clare (1971: 45) actually claimed that large areas of the reef were dead as a result;

3. The island has no supply of fresh water and the hotel has to use salt water in its septic tank system. The salinity inhibits the bacterial breakdown of sewage and, as a result, poorly treated wastes are being discharged into deeper water, later to be washed up onto the reef. Damage to marine life is inevitable;
4. Oil seepages from motorboats and the ferry linking the island with the Australian mainland are also suspected of affecting the coral and fish life but in what ways and to what degree have not been determined. Clare's examination is further evidence that the survival of ecosystems with low resilience is threatened by tourism.

Mountains

Mountains and upland areas have been besieged by people for the purposes of agriculture, forestry, reservoir development and, more recently, tourism and recreation. Mountains have attracted tourists for several centuries and they have also been the location of a large proportion of national and other park developments. With the emergence of mountain and upland recreational activities, such as skiing, climbing and some forms of hunting, these environments are being more intensively used. Previously remote and inaccessible areas are being opened to the influence of tourism.

Mountains contain a wealth of life zones which are squeezed into several thousand vertical feet, rather than across thousands of square miles. The layering of vegetation, or life zones, is reflected in the distribution of wildlife. Wildlife migrates between these life zones in response to seasonal and climatic fluctuations and the availability of food. The diversity of vegetation is of great importance to mountain wildlife for both food and protection. The disruption of these life zones by tourist activities can cause pronounced environmental effects. Tourist accommodation, ski lifts, cablecars, access roads, power lines and sewage systems are, in effect, squeezing the life zones into even more compact conditions and, on some occasions, even obliterating one or more of them. Animals often seek refuge from storms on the lower slopes. As the timber line is gradually pushed up the mountain the margin of survival for wildlife is greatly reduced.

The effects of tourist developments in mountains are numerous. Roadways alter drainage patterns and run-off from them can pollute streams. If they are built in previously inaccessible valleys they may prevent some animals from migrating and wintering there. Ricciuti (1976: 32) reported that this is now occurring in some areas of the Himalayas. He also noted that erosion and landslides have resulted. Ski lifts and trails increase the likelihood of contact between humans and wildlife.

Even the most remote mountains are not free from such impacts. Socher (1976: 388) recorded that in 1962 Nepal received just over 6,000 tourists but by 1975 this figure had risen to 100,000. Trails of litter now line the mountaineering and tramping routes on the lower slopes of Mount Everest and are especially concentrated around the campsites. The effects are compounded as successive groups of tourists use the same sites repeatedly. Sayer (1973: 744) reported similar problems in British upland parks.

The disruptions to mountain vegetation, soil stability and wildlife are particularly important because they induce snowballing effects of irreparable and often hazardous proportions. For example, Socher (1976: 388) noted increases in the number and scale of landslides and rockfalls in the Himalayas. In the high energy environments of mountains, where slopes are steep and climates are often extreme, the carrying capacity is likely to be small and environmental disruptions may take centuries to disappear (Willard and Marr 1970: 1971). Being both highly attractive and ecologically valuable, but of low resilience to the impacts of mass tourism, such special environments pose difficult but urgent questions for environmental protection.

Impacts of tourism on the man-made environment

So far it has been assumed that the basic resources of tourism are predominantly natural attractions and that risk to the environment is created primarily by excessive numbers of tourists using these resources. Natural attractions, by themselves, are insufficient to satisfy the tourist for they must be complemented by other tourist facilities and a supporting infrastructure. One of the most obvious environmental effects of tourism is the development of these facilities and infrastructure. They can be observed in three major types of tourist development:

1. The growth and form of tourist resorts;
2. Hotel development in cities;
3. Second home developments in rural environments.

These three types of development are discussed below and the problems which are associated with each are examined.

Tourist resorts

Growth and change in tourist resorts

Tourist resorts are not new landscape features. Peters (1969: 157) wrote that 'for generations spas, winter sports centres and coastal resorts have existed, developed around a few outstanding natural features and aimed at specialized markets'. Resorts were developed essentially to cater to the needs and desires of tourists. Resorts often developed from existing villages and towns, either changing the structure and spatial organization of the town itself, or growing in its immediate vicinity. Their locations were determined largely by the means of access. Limitations on travel restricted visitors to attractions which included, or were in close proximity to, accommodation and dining facilities. By the end of the nineteenth century four major types of resort had emerged (Lawson and Baud-Bovy 1977: 63). These were:

1. Spas for health and entertainment, e.g. Baden-Baden, Bath and Buxton;

2. Climatic resorts which existed for the treatment of tuberculosis and other diseases, e.g. Leysin and Menton;
3. Alpine resorts, e.g. Chamonix and Zermatt;
4. Seaside resorts for health cures and recreation, e.g. Bordighera, Brighton and Deauville.

In recent years many traditional resorts have suffered from declining patronage. This has resulted from changes in fashion, market structure, travel motivations and in tourists' choices of accommodation and recreation. The growth of new resorts which offer highly desirable and relatively reliable climates, new and exciting surroundings, and modern facilities has also been responsible for the declining appeal of many older resorts. Nevertheless, many older resorts have survived the effects of change and competition by introducing new attractions. This has been achieved through the rehabilitation of existing attractions and a reorientation towards new market opportunities. For example, the establishment of new points of interest has contributed to Brighton's continued tourist patronage and Atlantic City has attempted to revive its sagging economy through the introduction of gambling (Stansfield 1978).

Butler (1980) has suggested that resorts go through a cycle of evolution, passing through stages of exploration, involvement, development, consolidation, stagnation and either decline or rejuvenation. The exploration stage is characterized by small numbers of tourists who make individual travel arrangements and follow irregular visitation patterns. At this stage there are no specific facilities provided for visitors and the physical fabric and social milieu of the area are unchanged by tourism. As numbers of visitors increase and assume some regularity, some local residents enter the involvement stage and begin to provide facilities primarily, or even exclusively, for visitors. The development stage reflects a well-defined tourist market area shaped in part by heavy advertising. Some locally provided facilities disappear, being superseded by larger, more elaborate, and more up-to-date facilities, particularly accommodation. Changes in the physical appearance of the area are noticeable. At the consolidation stage major franchises and chains in the tourist industry will be represented but few, if any, additions will be made. Resort cities now have well-defined recreational business districts and old facilities may now be regarded as second-rate and far from desirable. As the area enters the stagnation phase, the peak number of visitors will have been reached. Natural and genuine cultural attractions will probably have been superseded by artificial ones. Capacity levels will have been exceeded, with attendant environmental, social and economic problems. The area will have a well established image but it will no longer be in fashion. In the decline stage the area will not be able to compete with newer attractions and so will face a declining market. Property turnover will be high and tourist facilities may be replaced by other structures as the area moves out of tourism. Hotels may become condominiums, convalescent or retirement homes, or conventional apartments, since the attractions of many tourist areas make them equally attractive for permanent settlement, particularly for the elderly. On the other hand, rejuvenation may occur, although it is unlikely that this will take place without a complete change in the attractions on which tourism is based. Clearly, each of these stages has different implications for the environment, as well as varied economic and social consequences. The latter topic will receive more detailed discussion in the next chapter.

The classification of resorts

Resorts are types of town which can be distinguished from other urban centres by their specialist functions. The literature often refers to resorts as coastal, inland, mountain or scenic on the basis of their geographical location. Robinson (1976: 168) has classified resorts according to the character of their development. He recognized two major categories:

1. Centres which have developed exclusively as tourist resorts either by adding artificial attractions and infrastructure to pre-existing natural attractions, or by developing tourist infrastructure in the absence of striking natural resoures. Blackpool, Monte Carlo and Niagara Falls are examples of such resorts;
2. Towns which have developed a tourist industry as an incidental part of their normal functions. Stratford-on-Avon, Stratford (Ontario), and most capital cities are examples of such places.

A third resort type could be added to Robinson's classification, although some might regard it as a specialized case of the first category. This is the recently developed, planned, integrated resort. All facilities and services are located within the resort in appropriate places and on the correct scale and design according to a preconceived plan. S'Agaró on the Spanish Costa Brava was one of the earliest examples of this type of development, although it has since been joined by such places as Cancun in Mexico, and the resorts of the Languedoc-Roussillon coast in the south of France.

The morphology of resorts

Resort development, by making an attraction suitable and more amenable for tourist consumption, typically transforms the natural environment. The concentration of tourist facilities and services is a clear example of an environmental transformation (Cohen 1972: 170). In the development of resorts, tourism has an urbanizing effect (Cohen 1978: 226). Resorts acquire a distinctive identity and unique morphology through the concentration of specialist facilities. This was observed by Jones (1933: 374) for Banff as early as 1933. The topic has received sporadic attention since that time. Early studies concentrated upon seaside resorts and, to a lesser extent, spas. In this respect, the early works of Gilbert (1939, 1949, 1954) and Barrett (1958) were especially noteworthy. Stansfield (1978) has considered the growth of Atlantic City and both he (1972) and Demars (1979) have drawn comparisons between the European and American experiences. Pigram (1977) has undertaken a similar study in Australia and Pearce (1979) has attempted to develop a model of tourist space, drawing upon Japanese and French precedents. Stansfield and Rickert (1970: 215), on the basis of an examination of Niagara Falls, Ontario, and coastal resorts in the United States, developed the concept of the recreational business district. Other studies by Lavery (1971), Robinson (1976) and Lawson and Baud-Bovy (1977) have discussed the forms and functions of resorts in a more general fashion. The latter authors extended their discussions beyond coastal resorts and drew comparisons between the forms and functions of resorts of different types. The relationships between transportation developments and resort morphology have also received attention (Stansfield 1971; Wall 1971, 1975). Thus, there is now a substantial, if widely scattered, literature on resort morphologies but, even so, the topic has not attracted attention to the same extent as that of other urban centres.

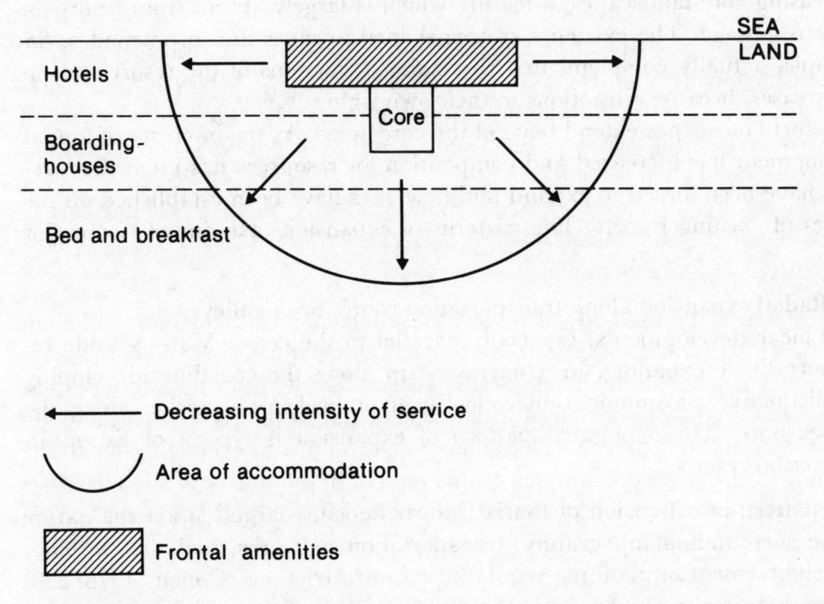

Fig. 7 Theoretical accommodation zones in a seaside resort (*source*: Wall 1971, after Barrett, 1958)

Tourist development usually begins around a core and expands outwards from that centre. In coastal resorts the core area of major shops, dining and entertainment facilities, large hotels and businesses is located in close proximity or adjacent to the main zone of frontal amenities. As the resorts grew a concentric pattern of architectural and social stratification developed. Hotels occupied prime locations around core facilities, while boarding-houses and bed and breakfast accommodation located less centrally. Barrett (1958), on the basis of an investigation of eighty coastal resorts in England and Wales, found that the intensity of accommodation services decreased with increased distance from the core region (Fig. 7). Railway development, which stimulated the growth of many European seaside resorts, created a characteristic 'T-shape' as a main street developed from the station to the linear shoreline promenade. Pearce (1978: 149) noted, somewhat similarly, that the older French resorts developed along the waterfront: 'Typically this consists of a parallel association of the beach, a promenade, a road or highway and a final line of accommodation and commerce where the best hotels and most expensive shops and apartments are to be found along with the casino.' The American coastal resort appears to differ slightly from the European in that the road runs behind the accommodation and does not separate it from the beach as it usually does in the European case.

The core area is usually the oldest and most intensively developed section of the resort. It is also the area which has undergone the greatest transformation. Core areas of resorts which have experienced a long history differ markedly from those of recent origin. Patmore (1968), examining spas and seaside resorts in Britain, noted that they occasionally developed aesthetically pleasing core landscapes, a feature which is largely absent from resorts of modern vintage. The existence of tree-shaded avenues and promenades, for example, actually complemented the major attractions of the resort and, in many cases, became attractions in their own right.

Resort landscapes extend beyond the core areas. As the intensity of tourist development has increased and competition for resources has intensified, resorts have been forced to expand and new ones have been established on the fringes of existing resorts. The patterns of expansion take one of two major forms:

1. Radial expansion along transportation routes or in valleys;
2. Linear developments, especially parallel to the coast. Many seaside resorts have expanded in a narrow strip along the coastline, developing distinctive accommodation zones in an unbroken succession along the seashore. This elongated pattern of expansion is typical of European coastal resorts.

The pattern of expansion of tourist resorts depends largely upon the nature of the surrounding topography, transportation networks, and the existence and enforcement of planning regulations and restrictions. Cohen (1978: 228) stated that, in the absence of such controls, much of the peripheral development of tourist resorts has been in areas with only secondary resource qualities and has led to the establishment of contrived attractions of unique appearance, such as discos, strip-tease bars and casinos.

Renewal and expansion of old resorts in Europe is a relatively recent trend which is not well documented in the literature. Pearce (1978: 151–2) illustrated this trend using the example of Biarritz, a well established and widely known resort on the Atlantic coast of France. In this case, as with many other older coastal resorts, the beach front sites are well developed making it essential not only to renew existing developments, but also to stimulate and concentrate further growth around another focal point of interest. Both developments are evident in Biarritz and also in the other French resort towns of Carnon and St Cyprien.

Environmental impacts of resort developments

The growth of resorts, with their distinctive morphologies developed solely for recreation and tourism, has been one of the most significant environmental transformations associated with tourism. However, the environmental impacts of recent resort developments have not received much detailed examination, although they have sometimes received discussion in the popular press. From the limited amount of research which has been undertaken the following impacts are the most prominent effects:

1. *Architectural pollution.* There has often been a failure to integrate resort infrastructure with aesthetically pleasing characteristics of the natural environment. Large, dominating hotel buildings are often out of scale and clash with their surroundings. Pearce (1978: 152) referred to this as 'architectural pollution' and it is the inevitable result of the juxtaposition of buildings in widely different architectural styles. The failure to incorporate adequately environmental considerations into the architectural designs of hotels and dining and entertainment facilities can lead to consequences which are both aesthetically unpleasant and economically unprofitable. Wimberly (1977: 26) stressed the economic benefits of architectural excellence when he wrote: 'In the case of a hotel, the attractiveness of the physical structure is vital to the facility's success. In a very real sense, design and ambience have a "dollar and cents value". Every hotel requires thoughtful design but resort hotels – which exist only for the pleasure of their users – demand it';
2. *Ribbon development and sprawl.* Coastal resort development, particularly in the absence of planning regulations or restraints, has tended to sprawl along the coastline. This is a response to the need to take advantage of the beach as a primary resource, and because of the availability of less expensive land for building. Ribbon development has also occurred along valleys and scenic routes in inland areas. In many cases the development has been of low quality and is left unoccupied for most of the year;
3. *Overloading of infrastructure.* In many resorts, infrastructures are unable to cope with the intensity of tourist visitation at peak periods of the year. The result is supply failures, pollution and health hazards;
4. *Segregation of local residents.* The spatial separation of tourist areas from the rest of the resort, or the surrounding countryside, makes for social segregation. The mass tourist may be surrounded by, but not integrated with, the host society. Separation is clearly seen in cases where tourists enjoy special facilities which are unavailable to residents, or where residents go to areas other than the core for their recreation or purchases of goods. This situation may be aggravated by the line of high-rise hotels which may constitute a physical barrier, both visual and real, between the inner residential zones and the prime attractions of the core area;
5. *Traffic congestion.* This has emerged as one of the more serious consequences of resort development. In a poll conducted by the Swiss Tourism Federation in various Swiss holiday resorts, visitors actually rated 'traffic chaos' as the greatest problem facing resorts and the one requiring the most urgent solution. An attitude study undertaken by Rothman (1978)

revealed that residents often feel the same way. An extensive study by Schaer (1978) described the emergence of traffic problems in resorts and discussed possible remedial actions. He noted that the traffic problem takes three forms:

(a) the mutual obstruction of different modes of traffic, particularly the conflict between pedestrians and motorcars;

(b) traffic overload and congestion at key points within resorts, especially at access points and in the main streets of the core area;

(c) the discrepancy between the demand for and availability of parking space. Parking space is required by excursionists in close proximity to points of attraction, by tourists at hotels, shops and restaurants, and also by local inhabitants at their places of residence and work. The lack of unused space in resorts and the high cost of providing such facilities have been major factors contributing to the imbalance between supply and demand.

Integrated resort developments

In principle, the creation of integrated resorts with fully planned developments should ensure that the objectives of the developing authority, the local population and the tourists are adequately met. The 'Disneyworld' type of self-contained project, as well as being profitable, is aimed at easing the burden on existing infrastructure and localizing social effects (Burn 1975: 28). The recently developed ski resorts of Cervinia in Italy and Alpe-d'Huez in France, and the Spanish coastal resort of S'Agaró on the Costa Brava are three of the better known fully planned resorts. However, Lawson and Baud-Bovy (1977: 65) were sceptical of their success in ameliorating many of the problems faced by other less planned resorts. They identified seven major failings of many integrated developments:

1. The resorts have become overdeveloped, incurring similar problems of traffic congestion, high-rise and high-density buildings, noise and pollution, and high costs and inflation, as less planned locations;
2. In taking advantage of prime locations, the resorts have emerged as dominant features in the landscape and have frequently modified and eroded the area's natural resources;
3. Speculative development still exists in both public and private projects;
4. Seasonality in use leaves many of the facilities vacant for a large proportion of the year;
5. There is little benefit to local inhabitants as most of the seasonal workers, food and other commodities are imported;
6. The resort separates the tourist from the host population and, occasionally, local inhabitants are prohibited from using tourist facilities;
7. Short-term planning goals are dominant. Short-term economic benefits to developers have often outweighed long-term social and economic benefits for both the visitors and their hosts.

In recognizing these failures the authors suggested that an appraisal of the successes and shortcomings of resorts, and an assessment of the planning process in general, should be undertaken before new resorts are developed.

Remedial measures to ameliorate resort development problems

Many resorts are currently formulating and implementing regulations and remedial measures to overcome some of the detrimental environmental impacts produced by the rapid growth of tourism in places like Hawaii, Jamaica and Spain (Crittendon 1975: 8). In Bali, for example, high-rise, western-style hotels will not receive planning approval in locations on the island which are in close proximity to unspoiled Balinese villages. Tunisia insists that new hotels conform to indigenous Moorish architectural styles. Adherence to local principles of architectural design and local construction materials are also required by planning authorities in other countries. The French hotels of La Grande Motte and the overlapping chalet development of Mont d'Arbois, and the Lobo Wildlife Lodge in Tanzania, are examples of aesthetically pleasing hotel designs. They represent attempts to integrate structures into the attractive natural surroundings and still maintain high occupancy densities.

Remedial measures to alleviate problems of traffic overloading have also been widely used. New resorts have adopted designs and policies which regulate the entry of cars in core areas and this has reduced traffic congestion, noise pollution, hazards and the high cost of road maintenance. Private vehicles are only permitted entry to load and unload passengers and luggage at accommodation points. Parking is limited to a car-park or garage at the point of entry. The French ski resort of Avoriaz, and Mackinac Island on Lake Huron, are examples of new resorts which ban private, motorized transport from core areas. Even more extreme measures have been introduced into Bali where international flights into Denpasar are limited. In Bermuda there are restrictions on the number of charter flights, size of motorcars, and the number of cruise ships permitted in the harbour at any one time.

Tourists visiting Hawaii's Waikiki Beach find luxurious high-rise hotels and lavish meals. However, they also find crowded beaches, congested streets, water pollution and contrived images of native life. Such situations have prompted many resorts to adopt similar remedial measures to those which have been described above. St Vincent and the Grenadines, which are small islands in the Lesser Antilles, are determined to avoid the mistakes made in Hawaii. Crittendon (1975: 12), reporting upon statements made by the islands' prime minister, indicated that mass tourism has been banned and the establishment of duty-free shops, swimming pools and casinos is prohibited.

It is difficult to determine the success of the various measures aimed at alleviating environmental problems because of the scarcity of evaluative studies of planned tourist projects. The closing of streets to vehicles and the building of pedestrian precincts are small steps forward in controlling tourist development and in reducing its environmental impacts. In spite of the many precautions and controls taken to impose restrictions on tourist development and to reduce environmental impacts, Cohen (1978: 230) felt that the pressures of mass tourism are so strong that it is doubtful if the efforts will be successful in the long run.

Infrastructural change in urban areas

So far this discussion has concentrated upon resort landscapes and their environmental consequences. Physical change induced by tourist development is also apparent in many urban areas (Hutchinson 1980; Wall and Sinnott 1980). Tourism has become an important activity in most of the world's large cities. As noted already not all environments can withstand the influx of tourists and large urban centres are no exception. However, large cities, because of their size, can absorb large numbers of tourists without obvious repercussions. Nevertheless, like any other system, urban centres have their carrying capacities and, from some studies, it appears that these have been exceeded in many European centres. Hall (1970: 445) claimed that: 'the age of mass tourism is the biggest single factor for change in the great capitals of Europe – and in many smaller historic cities too – in the last 30 years of this century'. Harrington (1971: 176) adopted a similar but more sceptical outlook on the effects of London's unregulated tourist boom when he claimed it 'threatened the quality of life in the city and indeed the tourist industry itself'.

The horizon of hotels which has emerged in many European cities, including London, is having regrettable consequences. Most of the negative side effects can be categorized under the heading of land use conflict. Firstly, although hotels are technically residential properties, they displace people. Hotels in London are a profitable means of urban renewal but, unfortunately, they are invading areas with some of the highest residential densities in the city. The magnitude of residential conversion was illustrated by Plummer (1971: 32) when he estimated that, in one borough alone, in 1971, outstanding tourist projects would entail the loss of 1,000 residential beds. In addition, hotels have invaded areas formerly devoted to specialized trading functions. Covent Garden and some of the dock areas have been deserted by their original functions which have been transferred to other areas of the city. Hotel encroachment has also occurred in many areas that are particularly rich in tourist attractions. These include the corner pubs, bistros and lively entertainment spots of the West End. Furthermore, major clusters of hotels have displaced residential developments in conservation areas. Hotel development in such areas has resulted in a considerable loss of amenity to remaining residents through traffic congestion and obstruction, lack of parking space, and increased noise from late-night tourist activity. Hotel developments have been uniform in design, lacking in both architectural interest and aesthetic quality. Many of London's new hotels are identical to those found in the major tourist resorts of Europe and the Caribbean, and would not look particularly out of place in these areas. The desertion of the waterfront and the displacement of high-density, low-cost, run-down housing is seen by some as an improvement. However, the undesirable social effects of residential displacement and employment losses have caused the desirability of these developments in London to be questioned.

Hotels of good design, which are located in the right place, are an asset to the urban environment. They can aid in the rehabilitation of historic areas and rejuvenate decaying parts of the city. They can also stimulate the creation of subsidiary centres outside of the traditional city core (Young 1973: 113), although Hall (1970: 445) has pointed out that developers are often reluctant to build hotels in such places and planners appear to lack the insight and power to encourage tourist development in them. The development of urban hotels, too often, has denied other activities the use of land in key locations and has resulted in undesirable environmental and social consequences.

The problems of hotel development and urban tourism have been widely recognized in the literature but solutions have been identified only infrequently (Hall 1970; Harrington 1971; Young 1973). Young (1973: 124) suggested a number of remedies which would require action at both national and local levels. At a national level, these were:

1. The reduction of the demand for urban tourism by reducing the scale of advertising and the use of promotional media. If the tourist industry is to be successful in the eyes of both hosts and guests, destinations should not be oversold;
2. To influence national policy such that the flow of tourists to particular regions is optimal, i.e. neither too high nor too low. It is also necessary to convince policy-makers that each region has physical and social capacities beyond which any increases in patronage are essentially counterproductive.

Additional solutions, which require implementation by local governments, could include:

1. Calculation of the maximum number of tourists that the infrastructure of the city can take. This number can then be converted into a corresponding number of accommodation units. This works well in theory but methodological problems exist which mean that the results of any calculations should be considered with caution. Mobile tourists and those staying with friends and relatives may be overlooked. In addition, many properties have been converted illegally into tourist accommodation;
2. The introduction of a tax on hotels and other tourist accommodation. This should increase the price of accommodation and can be used to balance demand with supply. Higher prices will discourage excess tourists and encourage them to select an alternative destination;
3. The removal of financial incentives for developers in areas which already have sufficient hotels;
4. The planned allocation of hotel space to areas within the city to avoid the over-concentration of hotels and their associated problems. This will also diffuse the benefits and costs of tourism more widely across the city;
5. The lifting of rent controls on apartments is another, although radical,

alternative in the London situation. The relatively low rents which Londoners pay under the fair rent scheme encourage the conversion of flats and houses to hotels because returns to investment are considerably higher per hotel room than from an apartment.

Land use conflict is particularly acute in urban areas where land is at a premium and there is pressure to increase the quantity of open space, to reduce housing densities, to build urban motorways, to construct new educational and medical institutions, and to meet a host of other needs. London is typical in this respect of many large urban areas and a series of compromises must be made.

> The opportunities are immense but so are some of the problems they create for public authorities. We are anxious that London should keep its place as one of the world's greatest tourist attractions. But care is needed to ensure that its power to draw visitors from all over the world does not produce a situation which is self destructive (Plummer 1971: 33).

Second home developments

Infrastructural changes resulting from tourism are highly visible in resort landscapes and in hotel developments in urban centres. The recent proliferation of second homes has been less obvious. Second residences may be used for long vacations, but weekend domestic tourism is most common. The demand for second homes is related to general recreational demand variables, to the degradation of urban environments, the increasing size of conurbations, and the opportunities for real estate investment (Lawson and Baud-Bovy 1977: 83).

The locations of second homes in Europe and North America indicate three major areas of preference:

1. Countryside areas within easy access to large urban centres;
2. The coastline;
3. Elevated regions with attractive and picturesque scenery (Clout 1971: 531).

These are areas which are vulnerable to deterioration from the pressures created by weekend and summer visitors.

In spite of the renovation and restoration of many rural cottages and the economic advantages induced by second home development, the countryside is generally considered to be worse off environmentally from this landscape change. A degraded condition may result from a number of causes:

1. The clearance of vegetation for road access and for construction disrupts wildlife and reduces soil stability;
2. The deposition of human wastes into rivers and lakes reduces water quality and is potentially dangerous to participants in water-based recreational activities;
3. If located in prime recreational environments, such as lake margins,

islands, river bends, the forest edge or on hillsides, their visibility may decrease the aesthetic value of that particular locality. There has been little concern for aesthetically harmonious designs evident in most second home developments.

The reader interested in gaining more information on second homes is referred to the collection of papers on the topic which has been edited by Coppock (1977).

Summary

In all of the above cases of environmental impact and landscape change, i.e. resorts, hotel encroachments in urban areas and second homes, the initial impact of development is the start of a succession of changes. Remedial measures to alleviate such secondary effects as traffic congestion, visual pollution and inflation, have met with only mixed success. Regulations and ameliorative measures should be applied directly to the cause of such problems, which is the tourist industry itself. In the absence of adequate legislation and planning controls, the development of tourist infrastructures has been left largely to the interplay of market forces. These have failed to ensure that adequate attention has been devoted to environmental and social concerns, and have induced conflicts between developers, local residents and tourists.

Tourism and competition for resources

Problems of conflicting interests are evident in a number of areas of tourist development. The needs of the visitor have to be reconciled with the requirements of the resident population in the tourist destination. It has already been pointed out that the encroachment of hotel complexes into urban areas has caused the relocation of residential areas and industries. Tourist regions are frequently areas of aesthetic beauty and some forms of economic activity may be incompatible with the maintenance of this asset. Archer (1973: 4) has suggested that even different forms of tourist development may be incompatible.

Tourism may exist simultaneously and often in conflict with other activities, or it may be the dominant economic activity, eliminating or suppressing other activities dependent upon the same resource. In earlier sections of this chapter mention was made of land use conflicts in urban environments (Hall 1970; Harrington 1971; Young 1973). Studies have also been conducted which stressed the role of tourism in stimulating preservation of the environment while, at the same time, placing additional pressures on traditional economic activities (Myers 1972; Rensberger 1977). Of the numerous conflicts associated with tourist developments, conflicts in rural landscapes have received most attention.

The growth of tourism and recreation in the countryside has been attributed to a number of factors:

1. The saturated conditions in many tourist resorts have stimulated tourists to seek alternative destinations;
2. The enactment of legislation promoting travel to country areas. For example, in Britain the National Parks and Access to the Countryside Act of 1968 facilitated the creation of 'country parks' and 'picnic sites', thereby making rural areas more accessible to tourists;
3. The particular attractions of rural environments and the growth of rural tourism. The attractions of rural environments for tourists fall into five main categories (Bracey 1970: 258; Robinson 1976: 180). They are:
   (a) the passive enjoyment of rural scenery;
   (b) the desire for open space, quiet and peace of mind;
   (c) the desire to partake in rural sports such as hunting and fishing;
   (d) the desire to view and experience ethnic attractions such as colourful folk life, customs, house types, foods, drinks and festivals. This is particularly popular in areas of Eastern Europe, Asia, Africa and Latin America where rural environments have not been absorbed by surrounding urban centres. Educational and historical attractions, such as castles, churches, monastic sites and other historic features also draw many tourists to rural areas;
   (e) the use of the country for second homes.

Rural tourism has also developed on farms which are marginal for farming and only provide a small income in agriculture for their owners. In order to escape from the economic uncertainties of marginal farming, a growing number of farmers have developed recreation and tourism enterprises on their properties. Farm vacations, and the use of facilities and services provided by farmers, have become popular in the United Kingdom, Ireland and, to a lesser extent, Canada in the past decade. It has been estimated that between 10,000 and 15,000 farms (4 to 5 per cent of the total) in England and Wales now operate recreation and tourism enterprises that are open to the public (Countryside Commission 1974: 1). A wide variety of enterprises contribute to the 'agro-tourist' product (Lawson and Baud-Bovy 1977: 83). These include farmhouse accommodation, self-catering flats and cottages, and camping and caravan sites; resource-based activities such as riding, fishing and shooting; and other attractions such as country parks, farm museums, handicraft displays and educational tours.

Most of the literature on rural tourism has emanated from the United Kingdom, largely through the efforts of the Countryside Commission. In the United Kingdom agro-tourism has emerged primarily through the initiative of rural communities and their recognition of the benefits that tourism can bring. Ironside (1971: 3), writing in the Canadian context, suggested two principal types of benefit:

1. The acquisition of additional income through the leasing of land, the rent or sale of buildings, the sale of produce, food and handicrafts, and from part-time family employment in local tourist activities and services. Revenues may also be gained from rents, fees and licences from individuals and clubs using the farm for such activities as hunting and

fishing, and from the reduction in local taxes by non-farm people purchasing property for seasonal use and thereby reducing the assessed tax value;

2. The derivation of protection from land-use conflicts. The control of land-use conflicts in locations close to urban areas is in the farmers' interests. The land in such areas may be catering to urban recreational and tourist demands, thereby relieving pressures to provide facilities and services within the city. As it is in the best interest of the city to maintain rural recreation areas, where legally possible, they may supply free conservation and management services, and protect the land from trespass, vandalism and careless hunting.

Considerable commitment of land, capital, labour and personal effort and sacrifice are required of farmers and their families if they are to cater to tourist demands effectively and gain financially from them. Buildings and land may require modifications or improvements; streams, ponds and woodland may need protection; and pathways and riding trails may have to be provided. As a result some farm income may have to be forgone. These are all taxing requirements that the farmer will have to accept, and he and his family will also have to tolerate greater workloads and the invasion of their private life (Dower 1973: 467).

Tourism may be an ancillary source of income as well as an alternative form of land use. It has also induced conflicts in the countryside. Most conflicts have arisen in unplanned and unmanaged situations, and have been accentuated by unimpeded public access and the lack of restrictions on the type, location and intensity of tourist activities (Cullington 1980). The most common conflicts are:

1. *Conflicts of access.* Tourists may require to pass through farms to points of scenic attraction;
2. *Irresponsible behaviour by tourists.* Damage and destruction of crops, farm buildings, and harassment of livestock are frequently reported. Fire, excessive noise, illegal hunting and fishing, and litter also lead to conflicts between farmers and tourists. Butler's (1978: 197) attitude study of farmers in the Isle of Skye revealed this conflict as a major concern of farmers and as a cause of their opposition to the presence of tourists;
3. *Competition for labour by the tourist industry.* Farm labour is often able to secure higher incomes from employment in the tourist industry and this can create labour shortages, especially at peak working times on farms;
4. *Increased erosion* from overuse of paths and trails, especially from horse riding along river banks and in areas with sensitive vegetation;
5. *Increased competition for land.* Urban dwellers may seek to rent or buy land for weekend residences and hobby farms. If intensification of existing land holdings is not viable, then the expansion of farm size is the remaining alternative for increasing production. Competition for land and escalating prices may prohibit this.

There is an even greater potential conflict between tourism and agriculture. Much agricultural land is of only average quality and a great deal is second-rate or poor. Fortunately, from the perspective of tourism, these lands often afford better natural scenery than much of the highest quality, flat, agricultural land. However, the future of much of this high-quality land may be in jeopardy. Urban areas are expanding on such lands, and the food demands of a rising population must be met by the increased production of crops and livestock. Demands for participation in recreation are also increasing and the energy crisis may encourage many people to try to satisfy these needs closer to home. Agriculture and recreation can be compatible in many cases, but trends towards increasing intensity of both agricultural and recreational use may lead to greater conflict in the future if the situation is not carefully managed.

Tourism also competes for space in rural environments and, on rare occasions, can even compete with industry. A recent study by Aiken and Moss (1976) revealed an unusual example of this conflict in peninsular Malaysia. A unique limestone formation, known as the Batu Caves, is the focus of a number of competing activities. It is a Hindu religious shrine, a tourist attraction, and the site of a cement works and quarrying activities. The caves house religious statues, photographic exhibits, and a rich flora and fauna which, along with the cave structure itself, are a unique attraction. The existence of the caves has been threatened by the expansion of quarrying. Blasting, the burning of vegetation, and dust from the nearby excavation are already affecting the cave environment.

Bracey (1970: 258) stressed the need for a balanced approach to land-use planning to regulate, control and direct pressures and, at the same time, ensure that aesthetic values and indigenous economic uses are not harmed. This is easier said than done for it is not easy to reconcile the requirements of general amenity, water supply, nature conservation, forestry, agriculture and tourism.

Conclusions

Considerable knowledge has accumulated concerning the physical impacts of tourism. However, much of this knowledge has been concentrated in specific aspects of impact with the result that a somewhat unbalanced picture of the physical effects of tourism has emerged. Quite a lot is known about resort morphologies and infrastructural changes in urban environments but much less is known about the impacts of tourism on the natural environment.

Tourism exhibits two contrasting relationships with the environment. A symbiotic relationship exists when the interacting sets of phenomena are mutually supportive. The roles of tourism in the creation of wildlife parks and the preservation of historic buildings are examples of this relationship. Tourism may also exist in conflict with the environment. The trampling of vegetation, the pollution of resort beaches, and the irresponsible behaviour of tourists disrupting the feeding and breeding habits of wildlife are examples of this conflict. The relationships of tourism with the natural environment are

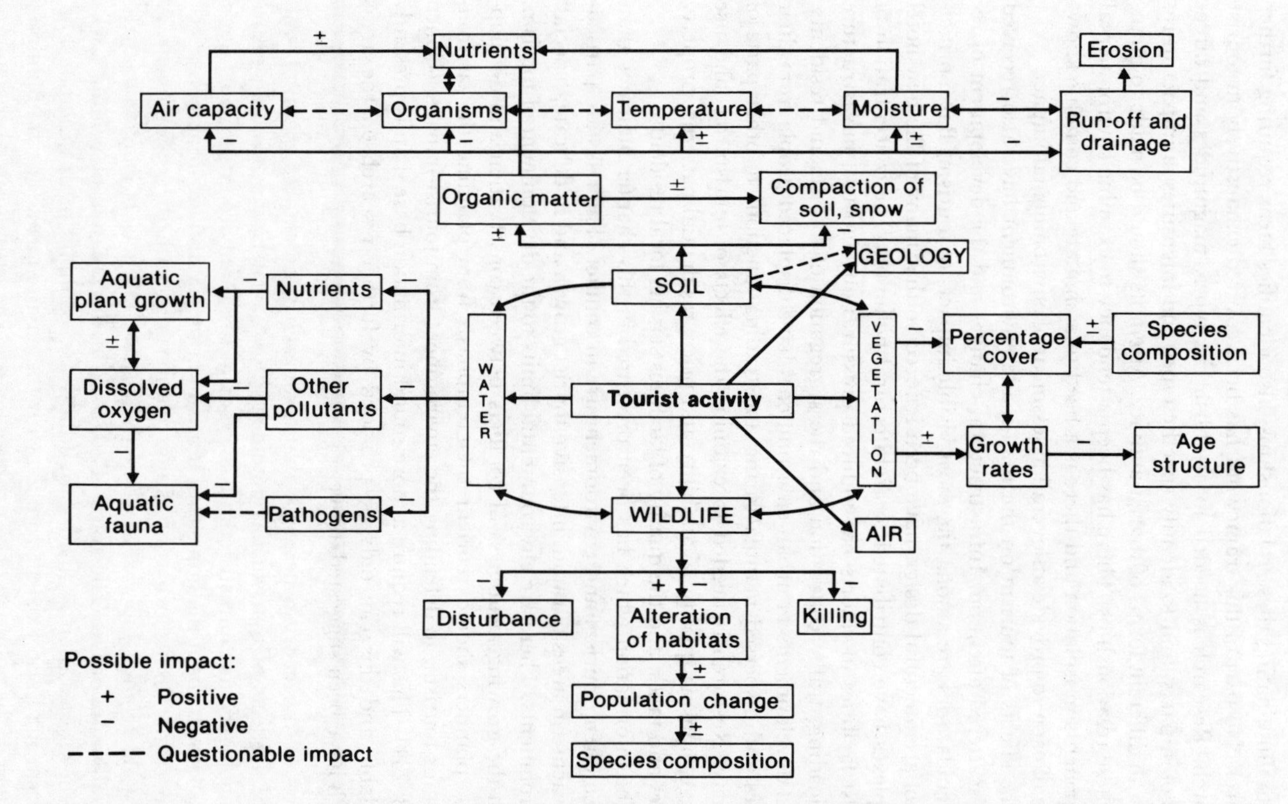

Fig. 8 Tourism and environment: impact interrelationships (*Source*: Wall and Wright 1977)

particularly ambivalent. This is a consequence of the complexity of the tourist phenomenon itself and the many interrelated components of the environment. The complex array of relationships between tourism and components of the natural environment is illustrated in Fig. 8. This framework, in addition to illuminating these relationships, also identifies areas requiring further research. No part of this framework has been covered exhaustively in tourist research. Research is needed to establish the types, magnitudes and directions of impacts, and to identify specific types and intensities of impact in relation to different forms of tourist activity. Attempts should be made to quantify these relationships. There has been a tendency to examine environmental components in isolation and there is a need to assimilate and combine knowledge of each to provide an overall assessment of environmental impact.

The effects of tourism on the man-made environment have also received only scattered attention. Infrastructural changes and the development of resort landscapes are among the most visible impacts of tourism. Basic principles of architectural design have been ignored so that many of the new hotel complexes have contributed to architectural pollution. The concentration of tourist facilities in resorts has induced pressures on land use, infrastructure overloading, traffic congestion and the segregation of tourists and residents. Hotel developments in urban environments have displaced people from their homes and dispersed manufacturing and trading functions to other parts of the city. Research is needed to examine the effects of tourism on land use change, housing quality, availability and price, residential patterns, property values and taxes, and the quality of facilities serving local residents.

The information which has been presented in this chapter indicates that, all too often, tourism and environment are in conflict. The risks are particularly acute in areas of rapid, intensive tourist growth and in delicate, 'special' environments. There are environmental limits to the development of tourism. With the growing awareness of conflicts between conservation and development, planners should reorient their emphasis from planning the environment for tourism, to defending the environment from tourist impacts (Cohen 1978: 234). This will require a more sound information base than is presently available and this can only be provided by further research on the areas which have been suggested above.

5 social impacts

Research on the social and cultural impacts of tourism falls into three different categories (Affeld 1975: 109). It has been concerned with:

1. *The tourist.* Such research has stressed the ramifications of the demand for tourist services and the motivations, attitudes and expectations of tourists;
2. *The host.* This research has emphasized aspects of the offering of tourist services. It is concerned with the inhabitants of the destination area, the labour engaged in providing services and the local organization of the tourist industry;
3. *Tourist–host interrelationships.* This research has been concerned with the nature of the contacts between hosts and guests and with the consequences of these contacts.

This chapter concentrates upon issues of the latter two categories of research and is devoted to the social and cultural implications to the inhabitants of destination areas which result from their relationships with tourists.

The social and cultural impacts of tourism are the ways in which tourism is contributing to changes in value systems, individual behaviour, family relationships, collective life styles, safety levels, moral conduct, creative expressions, traditional ceremonies and community organizations (Fox 1977: 27). In simplified terms, Wolf (1977: 3) stated that socio-cultural impacts are 'people impacts'; they are about the effects on the people of host communities of their direct and indirect associations with tourists.

Most of the early studies of the effects of tourism were restricted to economic analyses and enumerated the financial and employment benefits which accrued to destination areas as a result of the benefits of tourism (Pizam 1978: 8). In recent years a number of studies have emerged that examine the socio-cultural impacts of tourism. In contrast to the economic effects, such impacts are usually portrayed in the literature in a negative light. Some studies are very broad in coverage and have inventoried numerous impacts of tourism, ranging from the marketing of culture to the stimulation of racial, political and religious conflict (Young 1973; Jafari 1974; Turner and Ash 1975). Others have concentrated upon one particular type of impact (Jud 1975; Graburn 1976; Eadington 1978). A number of researchers have examined the socio-cultural impacts of tourism in a specific location, including the Pacific Islands (Farrell 1977; Finney and Watson 1977), Bali (Francillon 1975; McKean 1976), Spain (Greenwood 1972), East Africa (Ouma 1970), and in the Caribbean (Bryden 1973; Perez 1975; Lundberg 1974). De Kadt (1979) has provided a succinct statement on the topic which draws together the findings of other authors, some of whose research is presented in his book.

Recent research, in addition to illuminating some of the real adverse impacts of tourism on host communities, has also challenged the validity of the time-honoured belief that tourism promotes understanding between people of different nationalities and cultures. Peters' (1969: 126) contention that tourism creates 'social' benefits arising from the 'widening of people's interests generally, in world affairs and to a new understanding of foreigners and foreign tastes' is one of the few optimistic statements on the topic to be found in the relatively recent, major works.

In spite of the existence of the studies which have been mentioned above and a scattering of others, including Smith's (1977) collection of papers and Nettekoven's (1979, pp. 135–45) discussion of mechanisms of intercultural interaction, the socio-cultural effects of tourism on the people involved, i.e. those in tourist areas who become hosts and the tourists who become their temporary guests, remain an emerging area of investigation. The small, but growing, volume of literature which specifically considers the social impacts of tourism falls into one of two extreme types:

1. *Colourful stories*, usually in the consumer press (Apter 1974: 24). Jafari (1974: 242) quoted an example from a British humorist as follows:

   > What is the aim of all this travelling? Each nationality has its own different one. The Americans want to take photographs of themselves in: (a) Trafalgar Square with the pigeons, (b) in St Mark's Square, Venice, with the pigeons, and (c) in front of the Arc de Triomphe, in Paris, without the pigeons. The idea is simply to collect documentary proof that they have been there . . . . If you meet your next door neighbour in the High Street or at your front door you pretend not to see him or, at least, nod coolly; but if you meet him in Capri or Granada, you embrace him fondly and stand him a drink or two; and you may even discover that he is quite a nice chap after all and both of you might just as well have stayed at home;

2. *Specialized academic articles*. These studies appear in a wide variety of professional journals and books. Most of them are descriptive and they usually lack a strong theoretical base or analytical foundation.

The incisive studies of the latter group constitute the materials for this chapter. Although many studies of tourism make passing reference to the existence of social impacts, both positive and negative, most cast little light on the nature of these impacts or the means for their investigation. Such comments have received little attention here. The reports which have been given serious consideration are those which are either completely devoted to an examination of social impacts, or consistently refer to them in the course of the study. Many of these studies adopt a narrow focus, concentrating on a limited number of socio-cultural effects and ignoring or glossing over others. This is partly the result of what Marsh (1975b: 19) described as the 'incremental intangible costs' which are hard to measure and may be overlooked until major, irreversible changes in society or the environment occur. The need to adopt control or remedial measures has often arisen before such pertinent research issues as problems of measurement, the examination of processes of socio-cultural change, and the isolation of the determinants of tourist–host behaviour have been adequately resolved.

Tourist–host interrelationships

Socio-cultural impacts are the outcome of particular kinds of social relationships that occur between tourists and hosts as a result of their coming into contact. The nature of these relationships, or 'encounters', is a major factor influencing the extent to which understanding or misunderstanding is fostered by the tourism process.

According to de Kadt (1979: 50), tourist–host encounters occur in three main contexts: where the tourist is purchasing some good or service from the host, where the tourist and host find themselves side by side, for example, on a sandy beach or at a night-club performance, and where two parties come face to face with the object of exchanging information and ideas. He also pointed out that when it is claimed that tourism is an important mechanism for increasing international understanding, it is normally the third type of contact that is in mind. However, the first two types of contact are quantitatively more common. Nettekoven (1979: 135–6) noted that for mass tourists, intercultural encounters occur less during tourist travel than is often assumed and that intense encounters are less desired by tourists than is often suggested. Direct contact is not necessary for impacts to occur and the mere sight of tourists and their behaviour may induce behavioural changes on the part of permanent residents.

The tourist–host encounter, in which one or more visitors interact with one or more hosts, is staged within a network of goals and expectations (Sutton 1967: 220). On the one hand, the tourist is mobile, relaxed, free-spending, enjoying his leisure and absorbing the experience of being in a different place. In contrast the host is relatively stationary and, if employed in the tourist industry, spends a large proportion of the time catering to the needs and desires of visitors. Tourist–host relationships are also determined by the characteristics of the interacting groups or individuals and the conditions under which contact takes place. Most of the literature on tourist–host relationships is concerned with mass tourism. Under such conditions, the relationship is characterized by four major features (UNESCO 1976: 82):

1. Its transitory nature;
2. Temporal and spatial constraints;
3. Lack of spontaneity;
4. Unequal and unbalanced experiences.

Firstly, the relationship is transitory. A tourist's stay in one destination is usually short, ranging from a day or two if a vacation includes more than one destination, to three or four weeks, which is the normal length of a paid vacation. The temporary relationship is different for each of the interacting groups. Tourists often consider the meeting fascinating and unique because the host is frequently of a different nationality or culture. The hosts, on the other hand, may see it merely as one of many such superficial relationships which are experienced throughout the course of the tourist season. Boorstin (1961: 117) even claimed that, from the perspective of hosts, tourist contacts are nothing more than tautological experiences for residents of destination areas. As the tourist has a low customer loyalty, seldom returning to destinations more than once, the interaction between hosts and guests normally occurs only once and has little opportunity to progress beyond a superficial level. Where repeat visits occur, as in the case of some British vacationers, who return to the same boarding-house each year, a more intimate relationship may grow.

Secondly, the tourist–host relationship is characterized by temporal and spatial constraints which influence the duration and intensity of contact. The tourist often attempts to see and do as much as possible in the relatively short time available. As a result they may be more generous in their response to others, and more willing to spend money than they would be under more routine circumstances. On the other hand they may be unwarrantably irritable when even the slightest delay or break in travel plans occurs. Host reactions to the apparent urgency of the tourist to experience as much as possible in a short time eventually may become exploitative. As they continually provide tourists with simplified and condensed experiences of their area, hosts may develop a dual price and service system: one price and quality of service for the tourist and one for fellow residents.

Tourist reception facilities and services are frequently concentrated into a small number of complexes which are commonly referred to as 'tourist ghettos'. Tourism structures often reflect a desire, on the part of their owners and of local political authorities, for a certain degree of physical and social separation. By isolating the tourists and discouraging them from reaching beyond the tourist facilities, such organizations secure a greater proportion of tourists' spending money for themselves. It is conceivable that the only tourists who are motivated sufficiently to go beyond their immediate, artificial surroundings to mix with the resident population are the 'explorers' and 'drifters' (Cohen 1972: 168). Mass tourists, on the other hand, are controlled in their movements directly by tour operators or indirectly through the location of required services, such as accommodation, restaurants, entertainment and recreation facilities. In consequence, as tourism becomes more highly developed and resort areas expand, contacts between the majority of tourists and their hosts either do not take place or are largely infrequent and superficial.

Thirdly, tourist–host relationships lack spontaneity. Tourism brings certain informal and traditional human relations into the area of economic activity, turning acts of once spontaneous hospitality into commercial transactions (de Kadt 1979: 14). Package tours, planned attractions, exhibitions and all of their organized features are advertised and promoted well in advance of their actual occurrence. In effect, tourists trade off the prospect of convenient, comfortable and risk-free experiences that planned and organized travel provides, for less frequent and spontaneous contacts with their hosts. Meetings are more rigidly controlled and, at the extreme, may become a series of cash-generating activities.

Fourthly, there is a tendency for tourist–host relationships to be unequal and unbalanced in character. Material inequality often exists and is seen in tourist spending and attitudes. Hosts often feel inferior and, to compensate for this, exploit the tourists' apparent wealth. There are also inequalities in levels of satisfaction and the sense of novelty derived from the relationship. A vacation for the tourist is a novel experience but its consequences for the host are routine. Maintenance of that routine may be difficult with the advent of late plane arrivals and other travel emergencies. These may induce psychological pressures for the hosts to which the tourist is oblivious.

The literature which examines the socio-cultural impacts of tourism has usually been directed towards either social or cultural aspects. Using these terms very loosely, the social studies usually consider interpersonal relations, moral conduct, religion, language and health, whereas the cultural studies consider both material and non-material forms of culture and processes of cultural change. There is no clear distinction between social and cultural phenomena but this dichotomy has proved useful in categorizing studies and organizing the remainder of this chapter. A discussion of the social impacts of tourism will be followed by an examination of its cultural consequences.

Social impacts of tourism

The social impacts of tourism are not to be confused with the increasingly popular term 'social tourism'. Social tourism has not reached a high level of development in North America, but it has achieved more widespread acclaim in Europe. The objective of social tourism is to ensure that tourism is accessible to all people. Special efforts are made to include members of society who otherwise would be prevented from participating in tourist travel for reasons such as economic hardship, or physical and psychological disabilities (Kaspar 1977: 19). Social impacts of tourism may arise as a result of social tourism, but they are not endemic to it. Social impacts of tourism refer to the changes in the quality of life of residents of tourist destinations.

Framework for the measurement of social impacts

Few studies suggest ways of assessing the social impacts of tourism. Although such authors as Cohen (1972: 164), Robinson (1976: 137) and Pizam (1978: 8) have stressed the need for a greater emphasis on behavioural aspects of the visitor and resident, and the reciprocal effects of interaction between a community and its tourists, only two frameworks have emerged which appear to be widely applicable to social impact research in tourism. These frameworks constitute the beginning of the development of a conceptual basis for the assessment of social impacts. Both frameworks recognize that the social impacts of tourism change through time in response to structural changes in the tourist industry, and the extent and duration of the exposure of the host population to tourist development.

The first framework which we shall consider was developed by Doxey (1976) following his research projects which were undertaken in Barbados in the West Indies, and Niagara-on-the-Lake in Ontario, Canada. Doxey

(1975: 195) suggested that the existence of reciprocating impacts between outsiders and residents may be converted into varying degrees of resident irritation. Irritations may have their origins in the number of tourists and the threats which they pose to the way of life of permanent residents. Doxey argued that the responses of residents in different destinations will vary and that resident responses will change through time in a predictable sequence comparable to that of the resort cycle described by Butler (1980), Noronha (1976), Plog (1977) and Stansfield (1978). The value system of the destination is at the root of Doxey's framework and it is this which he considered to be of paramount importance in the measurement of social impacts.

Drawing upon his experience in the Barbados and Niagara-on-the-Lake studies, Doxey devised an irritation index (Table 16). Tourist destinations pass successively through stages of euphoria, apathy, irritation, and antagonism, to the final stage in which people have forgotten what they cherished and the environment is destroyed. The level of irritation arising from contacts between the hosts and the tourists will be determined by the mutual

Table 16 Index of tourist irritation

1. The level of euphoria
People are enthusiastic and thrilled by tourist development. They welcome the stranger and there is a mutual feeling of satisfaction. There are opportunities for locals and money flows in along with the tourist.

2. The level of apathy
As the industry expands people begin to take the tourist for granted. He rapidly becomes a target for profit-taking and contact on the personal plane begins to become more formal.

3. The level of irritation
This will begin when the industry is nearing the saturation point or is allowed to pass a level at which the locals cannot handle the numbers without expansion of facilities.

4. The level of antagonism
The irritations have become more overt. People now see the tourist as the harbinger of all that is bad. 'Taxes have gone up because of the tourists.' 'They have no respect for property.' 'They have corrupted our youth.' 'They are bent on destroying all that is fine in our town.' Mutual politeness has now given way to antagonism and the tourist is 'ripped off'.

5. The final level
All this while people have forgotten that what they cherished in the first place was what drew the tourist, but in the wild scramble to develop they overlooked this and allowed the environment to change. What they now must learn to live with is the fact that their ecosystem will never be the same again. They might still be able to draw tourists but of a very different type from those they so happily welcomed in early years. If the destination is large enough to cope with mass tourism it will continue to thrive.

Source: Doxey 1976: 26–7

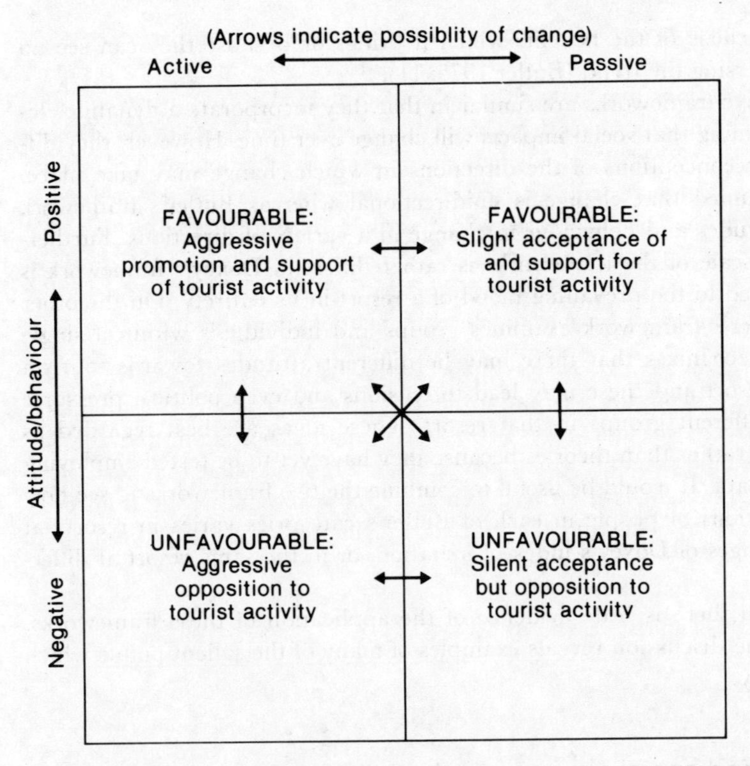

Fig. 9 Host attitudinal/behavioural responses to tourist activity (*Source*: After Bjorklund and Philbrick 1972: 8. Found in Butler 1974: 12)

compatability of each, with the assumption that even with seemingly compatible groups, sheer numbers may generate tensions, with differences in colour, culture, economic status and nationality as complicating factors. In Doxey's example of Niagara-on-the-Lake, it was clear that the number of visitors and the threat which they posed to lifestyles in the town, were a source of irritation which had reached serious proportions.

A framework developed by Bjorklund and Philbrick (1972: 8; 1975: 68) to analyse the processes which take place when two or more culture groups interact has been suggested by Butler (1975a: 89) as being applicable to the social relationships between tourists and their hosts. Figure 9 shows the matrix which is the basis of this framework. The attitudes and behaviour of groups or individuals to tourism may be either positive or negative, and active or passive, respectively. The resulting combinations of reactions to tourism may take one of four forms as shown in the diagram. Within any community, all four forms may exist at any one time but the number of people in any one category need not remain constant. For example, entrepreneurs who are financially involved in tourism are likely to be engaged in aggressive promotion while an often small but highly vocal group, uninvolved in tourism, can be expected to lead aggressive opposition to tourist development and the changes which it brings. The majority of the population is likely to fall into the two remaining passive categories, either silently accepting tourism and its impacts because of the benefits which it brings or because they can see no way of reversing the trend (Butler 1975: 11).

The above frameworks are similar in that they incorporate a dynamic element, assuming that social impacts will change over time. However, they differ in their conceptions of the directions in which change may take place. Doxey assumed that change is unidirectional whereas Butler's framework allows attitudes and behaviour to change in a variety of directions. Furthermore, the scale of the frameworks is rather different. Doxey's framework is to be applied to the prevailing mood of a resort in its entirety. On the other hand, Butler's framework examines groups and individuals within a single resort. It recognizes that there may be different attitudes towards tourism within a resort and these may lead to tensions and even political pressures between different groups in that resort. These ideas are best regarded as hypotheses rather than theories because they have yet to be tested employing empirical data. It would be useful to combine the two frameworks to see how the proportions of people in each of Butler's categories varies in resorts at different stages of Doxey's index of irritation, or in the same resort at different times.

Although there is little evidence of the application of these frameworks, the following discussion reveals examples of many of the salient points which they portray.

Tourism and social change: euphoria to xenophobia

The initial stages of tourist development are normally accompanied by enthusiastic responses on the part of individuals within the host population as they perceive the potential benefits that investors and visitors will bring to their area. As we have already seen, tourism can often bring new, and sometimes necessary, sources of capital and income that may supplement or replace traditional sources of earnings. For this reason tourism development has often received support from governments and local residents in stagnating or developing areas who recognize the economic benefits that may be earned. The initial euphoria and enthusiasm which are associated with the preliminary phases of tourism begin to dissipate as the industry expands and tourist numbers increase. As Fox (1977) noted in his review of social impacts in the Pacific Islands, the political leaders of newly developed destinations such as Tonga, the Cook Islands, Samoa and Fiji, at one time professed that: 'tourism will improve our country's economy and will benefit our island's people. . . . We are proud to have tourists see our culture and our beautiful island.' They now express a fearful concern for the increased strains imposed on traditional customs and lifestyles. Fiji's Minister for Communication, Works and Tourism epitomized local and governmental feelings towards further tourist expansion when he stated (Fox 1977: 31): 'Equally grave is the deterioration of social values in Fijian society . . . . We recognize some erosion of our traditional social fabric is inevitable in the pursuit of economic prosperity and we are prepared to accept a degree of erosion. What we are not prepared to accept is destruction of that social fabric.'

Host perceptions of and attitudes towards continued expansion of facilities and services to satisfy tourist demands may rapidly become increasingly antagonistic, and may eventually reach xenophobic proportions. Xenophobia occurs when the carrying capacity, or saturation point, is reached and exceeded. It refers to the overtly expressed resentment of and contempt for tourists and their behaviour. Attitudes of this kind vary in intensity and mode of expression from grieving statements expressed by hosts, such as: 'Many of our villagers are now in the towns trying to be like the foreign tourists that are swarming over us' (Fox 1977: 38) to outbursts of hostilities and violence such as those noted by Rivers (1973a: 250):

> Among black Jamaicans there is a growing consciousness that the kind of tourist industry which has sprung up around them is both demeaning and exploitative and resentment to white visitors is widespread. Trinidad has only just lifted its state of emergency; Antigua is playing down recent bomb outrages; the Bahamas still experience bouts of black militancy; and the United States Virgin Islands and Puerto Rico have rushed through costly courtesy campaigns to persuade the natives that tourists are good for them.

It is widely agreed, then, that in respect to socio-cultural impacts, there is a threshold of tolerance of tourists by hosts which varies both spatially and temporally. As long as the numbers of tourists and their cumulative impacts remain below this critical level, and economic impacts continue to be positive, the presence of tourists in destination areas is usually accepted and welcomed by the majority of the host population. Once the threshold has been exceeded, numerous negative symptoms of discontent make their appearance, ranging from mild apathy and irritation to extreme xenophobia, and from grudging courtesy to open exploitation. The critical point of tolerance varies between host groups with:

1. The cultural and economic distances between tourists and hosts. Resorts vary in their tolerance to peoples who differ from their own residents, especially by virtue of appearance, affluence, race and nationality. The greater the divergence of characteristics between the interacting groups, the more pronounced are the social impacts;
2. The capability of the destination and its population to physically and psychologically absorb tourist arrivals without undermining or squeezing out desirable local activities. The scale of dislocation is related to the travel intensity index, that is the ratio of visitors to residents (Lundberg 1974: 85). For example, in London, Paris and New York, cosmopolitan cities which are inundated by millions of tourists every year, tourists are absorbed by the vast complex of facilities and merge with their large resident populations. On the other hand, in Barbados and the Virgin Islands, tourist arrivals greatly exceed the size of the local population and overwhelm their facilities and services. Host resentment is likely to be more pronounced, therefore, in Barbados than in London;
3. The rapidity and intensity of tourist development. When tourism is introduced gradually into a large, established economy, the waves of im-

pact are usually small. Most developed countries, with the exception of small pockets of intensive development, have adjusted to the growth of tourism in the course of their general economic expansion. However, when tourism replaces another activity in a short period of time and heavy reliance is placed upon it as a dominant means of acquiring foreign exchange, socio-cultural and psychological repercussions are inevitable. Tourist resorts in the Greek Islands, the Pacific and the Caribbean, which are heavily committed to the tourist industry as a source of income, are incurring numerous unwanted social and cultural side-effects as a result of the rapid growth of the industry.

Although it is extremely difficult to determine if communities have reached the critical tolerance level, it appears that most studies have concentrated on areas either on the verge of, or currently exceeding, the threshold of tolerance.

Growing animosity to tourism, as Rivers (1973a,b) inferred, is not confined to the socially and economically deprived, but has also been expressed by concerned politicians, academics, churchmen, and even some businessmen. Past reluctance to mention the shortcomings of tourism because of its economic benefits is disappearing. Resentment tends to be highest in what Jafari (1974: 246) termed 'tourist magnetic' areas, where tourism is the principal source of income to the community, and all activities become oriented to accommodating tourist demand, which may be limited to a short season. Although the livelihood of residents may be derived from the presence of tourists, they view the approaching season with mixed feelings, and value the off season when only permanent residents are present (Jordan 1980).

Numerous situations provoke feelings of tourist resentment. The most intense feelings appear to develop from particular conditions:

1. *The physical presence of tourists* in the destination area, especially if they are in large groups. Residents frequently resent having to share facilities and services with visitors and often mention congestion as a problem;
2. *The demonstration effect*. Residents frequently resent the apparent material superiority of visitors and may try to copy their behaviours and spending patterns;
3. *Foreign ownership and employment*. The employment of non-locals in managerial and professional occupations, carrying greater responsibility and superior salaries to those available to local residents, also provokes resentment. These features have been labelled in the literature as outgrowths of neo-colonialism.

These conditions are examined in detail below.

The physical presence of tourists

There is ample literary and pictorial evidence which indicates the congregation of large numbers of tourists in resorts. The 'Jam-up at Vacation Spots' (US News 1976: 26) occurring on the beaches of Hawaii, in the ski gondolas of the Swiss Alps, in the harbours of the Virgin Islands and on the streets of Disneyland are expected in the tourist season and are even a sign of their success. Many studies have mentioned the presence of large numbers of tourists in particular places at specific times and the associated congestion of facilities and services. Few studies have attempted to estimate empirically existing levels of congestion or saturation. A number of crude measures have been employed as a first step towards such estimates. These include the ratio of the number of visitors to the local population, the average area of arable land per visitor, and measurements of the physical size of tourist facilities (Lundgren 1973: 2). Little is known about the responses of visitors to various levels of congestion as indicated by these measures, or the means which are available to alleviate the irritations which may occur with high levels of congestion. Tourist densities can be reduced by dispersing tourists more widely so that the ratio of tourists to hosts is reduced without decreasing the number of tourist arrivals. Alternatively, it may be feasible to create a tourist enclave on the theory that residents are less likely to be upset if their contacts with tourists are limited (Gray 1974: 394).

The demonstration effect

The disruptive role of tourism in reinforcing locally unattainable socio-economic aspirations has been noted frequently (Rivers 1973a: 250; Jafari 1974: 249–50). The introduction of foreign ideologies and ways of life into societies that have not been exposed to tourist lifestyles has 'tended to call up that all-embracing concept, the demonstration effect' (Bryden 1973: 95). The demonstration effect can be advantageous if it encourages people to adapt or work for things they lack. More commonly, it is detrimental and most authors indicate concern for the effects of foreign domination of the industry and the impacts of tourists who parade symbols of their affluence to interested hosts. Alien commodities are rarely desired prior to their introduction into host communities and, for most residents of destination areas in the developing world, such commodities 'remain tantalisingly beyond reach' (Rivers 1973a: 250). As a result discontent grows among the hosts. The hosts' resentment is heightened by the development of luxurious hotels and other 'foreign' tourist facilities, and also by unusual tourist behaviour. Tourists on vacation have fewer constraints than they do at home and they behave and spend in a less inhibited fashion. As a consequence hosts often develop misconceptions about tourists. Rivers, for example, reported that young Spaniards were convinced that all unattached female tourists were easy conquests. Similarly, Cohen (1971: 225) reported that fair-haired girls from Scandinavia were thought to be seeking sexual adventures in their travels and were sought by his study group of Arab boys. Furthermore, the hosts quickly perceive the desire of tourists to spend money lavishly to gain experiences and acquire souvenirs of their stay. UNESCO (1976: 93) suggested that, at the outset, hosts may develop an inferiority complex which sets off a process of imitation. The weaknesses of the tourists are quickly perceived and are subsequently exploited. A dual pricing system often develops and bargaining may be required for the purchase of goods and services. Sutton (1967: 219) portrayed this situation when he quoted: 'The American noticing the successive price reductions of the shopkeepers, the suppliant attitude and emotional appeals that may be merely routine for him, will perceive these as overdrawn and see the merchant merely as one skilled in exploiting his ignorance, playing upon his sympathies and fleecing his pocketbook.' It would not take many such experiences to fix an adverse stereotype in the mind of either host or guest.

Heightened economic expectations among the local population who aspire to the material standards and values of the tourists may lead to the copying of consumption patterns. In an attempt to attain the consumption levels of tourists, a growing number of the indigenous population may take jobs in service industries since these appear to offer a greater chance of advancement than traditional agriculture. Also an increasing proportion of the local population may try to migrate to other countries for fuller participation in western, cash economies.

Young members of host communities are particularly susceptible to the demonstration effect. Young people of traditionally closed societies observe the freedom and material superiority of young travelling westerners and respond in one of the ways discussed above to achieve similar status.

Most case studies which examine the demonstration effect agree that tourism can modify local behaviour and divide the population of destination areas. In her analysis of the Eskimo communities of Kotzebue and Nome, Smith (1977: 68) noted the development of a specialist minority population in the community. This group included those individuals who danced and demonstrated crafts to visiting tourists. They were mainly older members of the community who had retained, overtly, their traditional costumes and crafts and were, thus, in a position to capitalize on them. The young and better educated Eskimos acquired jobs in government and business in an attempt to achieve their western aspirations. Smith (1977: 71) described the older members of the community as 'marginal men' who have adopted some foreign ways of life, including new products, but, at the same time, live the culture of their ancestors.

Tourism has also changed the economic and social structure of rural communities in western nations. Greenwood (1976) and Pacione (1977) wrote on the effects of tourism on rural communities in Spain. In both cases traditional society resembled that of other rural Mediterranean areas consisting largely of fishermen, craftsmen, shopkeepers and peasant farmers. The advent of tourism had profound effects on the traditional peasant economies. Employment in the tourist industry has been one way for locals to improve their standard of living. Both Greenwood and Pacione noted that greater wealth and upward social mobility had become increasingly desirable, particularly for young members of the host communities. In consequence, class distinctions within the host communities were accentuated and reflected the degree of involvement in tourism. Moore (1970), working in the Canary Islands, showed that such changes could modify local political powers, with the emergence of new élites based upon tourism. These studies all show that tourism has improved the economic well-being for those involved in it, but it has also caused a transformation of the local communities.

Migration, both international and intra-national, is another important manifestation of the demonstration effect. Rural–urban migration is not a recent process nor is tourism a major cause of the phenomenon. Nevertheless, the ability of tourism to create employment, commonly in urban resorts, has facilitated the drift of people from rural to urban locations in many countries. Somewhat similarly, although on a different scale, aspirations of upward social mobility and higher material standards has caused many members of host communities to travel to foreign countries in search of employment and to satisfy these goals. The relationships of both intra-national and international migration to tourism development are not well documented but the example of Pacific Islanders migrating to New Zealand is enlightening. Since 1971 thousands of Tongans have migrated to New Zealand in search of employment. The extent to which either the unstable social and economic conditions in Tonga or the increased prosperity available in New Zealand has been the principal causative factor of migration has been debated by Urbanowicz (1977a,b). The natural resources of Tonga are inadequate to feed the growing population. In spite of attempts to increase and modernize the agricultural base, in an attempt to generate cash flows to meet consumer demands for western products, Tonga has had to export much of its own agricultural produce. Inflation and unemployment have placed great pressure on local resources and have encouraged the government to expand the tourist industry in an attempt to rejuvenate the economy. Tongans are migrating in response to the serious lack of domestic employment. Tourism may help to support the island's growing population but it may also aggravate other problems. The exposure to and subsequent demand for western products can be attributed, primarily, to the influence of tourism. Migration is contributing to social problems in both Tonga and New Zealand. In the latter country, racial tension and violence have emerged, and alcoholism and crime have increased. In Tonga a gradual erosion of the traditional extended family system, chain migration (continued migration aided by Tongans already in New Zealand) and an increased reliance upon remittance payments have become common (Koea 1977: 68–9). Although it would be wrong to ascribe all these ills solely to tourism, tourism has certainly played a part. In spite of the generation of employment and income, tourism has had serious effects on Tongan communities which include:

1. Increases in crime rates and tension with the inability of local residents to achieve their western-induced desires by legitimate means;
2. Intensification of resentments with the continued employment of expatriates in tourist occupations;
3. The disruption of traditional kinship and community bonds through the loss of important individuals by migration.

The secondary implications of the demonstration effect on employment and migration contribute to modifications in the internal structure of host communities. This may occur through:

1. Changes in the roles of women;
2. Changes in community cohesion;
3. Changes in demographic structure;
4. Changes in institutional structures and membership.

Information on these trends is fragmentary. In most traditional societies relations between the generations are governed by strict patterns of authority, underpinned by the financial dependence of youth on the older generation. The widening of employment and earning opportunities for both young people and women decrease their dependence and cannot but strain intrafamilial relations (de Kadt 1979: 43). However, it would be wrong to regard all such changes as undesirable. For example, Reynoso y Valle and de Regt (1979: 130–1) found significant changes in the employment of women in their Mexican case study. As a result of the opening of alternative job opportunities in new hotels, working conditions and salaries for domestics improved sharply. Opportunities also increased in the informal labour market for such activities as washing clothes, renting rooms, selling food and petty vending. Some women were able to formalize their businesses by opening hotels, restaurants and shops. According to Wilson (1979: 205–36), in the Seychelles females have received preference over males in a number of hotel and restaurant jobs, with the result that women lead less restricted lives than was previously the case. According to Wilson, 'If a young girl lands a good job, she can earn more money than her father and is better able to dictate her own life-style' (p. 224). However, in Tunisia, employers in the tourism industry have been prevailed upon to pay girls' wages to their fathers rather than directly to the employee (de Kadt 1979: 44).

The disruption of kinship and community bonds will reduce cohesiveness within the host society. This may be manifested in declining memberships in community organizations and a decrease in the quality of local services. However, Manning (1979: 157–76) has argued a contrary case and has suggested that tourism has contributed to a revitalization of black clubs in Bermuda. Greenwood (1976: 139) noted that the upward social mobility associated with the demonstration effect also implies residential mobility. In his Spanish study, he argued that the decline in community cohesion was most visible in the isolation and abandonment of the aged members of the community, many of whom were left alone and uncared for. The construction of tourist facilities can lead to an influx of labour with consequent changes in the demographic structure and associated social problems (Reynoso y Valle and de Regt 1979: 127–9; Tourism and Recreation Research Unit, University of Edinburgh, 1976–77). Again the situation is complicated for there are other cases where tourism has provided jobs, slowed the emigration of young people, and made a positive contribution to demographic structure (Diem 1980).

The undesirable consequences of the demonstration effect for the socio-economic development of host communities has been widely recognized. Turner and Ash (1975: 197) even postulated that the demonstration effect is tourism's chief weapon against the maintenance of authenticity in culture and community identity. They see tourism as an exporter of western lifestyles to developing countries, disseminating metropolitan values and decadence which are, themselves, being questioned in their countries of origin. The positive effects of demonstration are seldom mentioned, except by those involved in the promotion and marketing of tourism. It still remains to be determined if the display of a higher standard of living will produce an energizing effect on local people, encouraging more strict work ethics and increased initiative in search of wealth.

Bryden (1973: 96) adopted a different perspective and called into question the utility of the demonstration effect as an aid to explanation, calling it 'a vague, unsatisfactory concept'. He suggested that, on its own, the concept does not explain who is demonstrating what to whom; why, to what extent, or at what speed demonstration is occurring. He also asked why it is assumed that the tastes of foreigners are adopted by indigenes and not vice versa.

Tourism has undoubtedly engendered a more 'foreign' atmosphere in tourist resorts in both developed and developing countries. Radical viewpoints criticize the foreign domination of tourist services and facilities, and see beyond the demonstration effect a more insidious phenomenon which they attribute to neo-colonialism. This is another condition which promotes resentment of tourist activities.

Neo-colonialism

The view that tourism is a new form of colonialism and imperialism is one extreme perception of tourist development and its effects (Matthews 1977: 21). However, the evidence presented in chapter 3 and to be found in the writings of Levitt and Gulati (1970), Lewis (1972), Bryden (1973) and Perez (1975) indicates that this perspective is not without foundation. The movement of metropolitan citizens from the developed economies of Europe and North America to societies of the less developed world has a long history. According to some authors, the growth of tourism in these destinations has been only a change in the form and magnitude of travel without a major alteration in its colonial quality. Although the legal ties between metropolitan powers and tourist destinations has changed as many developing countries have received independence, the economic relationships between them have remained essentially the same. This condition has prompted the charge that tourism is a neo-colonial activity.

Three economic conditions substantiate this claim. Firstly, many developing countries have grown to depend upon tourism as a means of securing revenue. Many countries attempting to boost foreign exchange earnings have turned to tourism as a means of achieving this goal. Their success depends, in part, on their willingness to accommodate fully the needs of tourists. The West Indies, for example, waived tax and import restrictions, developed appropriate infrastructures and relaxed environmental safeguards in an attempt to encourage the development of the tourist industry. In other words, their political and economic priorities and organization have been directed towards the satisfaction of the demands of tourists.

Secondly, the development of tourism may be accompanied by a one-way transfer of wealth from the destination area to points of tourist generation. A large proportion of expenditures and profits flow back to foreign investors. High leakages may leave little profit in the destination area. A large proportion of the goods and services which are consumed by the tourist are produced at the place of origin and, thus, most of the profits are transferred to these points.

Thirdly, the employment of non-locals in professional and managerial positions and the frequent existence of foreign, absentee employers have also provoked charges of neo-colonialism. These features also contribute to high leakages through the remittances of salaries and profits to the tourist generating countries. Bugnicourt (1977: 2), writing specifically on African tourism, believed that these situations contributed to what he called 'tourism with no return'.

Perez (1975: 141), Bugnicourt (1977: 3) and Matthews (1977: 23) also argued that tourism is neo-colonialist through the herding of local groups into 'reservations' for the purpose of entertaining tourists, and through the transformation of the relics of past colonial regimes, such as old fortresses and historic buildings, into tourist attractions. Similarly, cultural symbols of art, music, dance and literature are exploited to cater to tourist curiosity and to make money.

The above discussion represents a radical perspective on the role of tourism in developing countries. There is certainly evidence to suggest that tourism is exploitive and that it displays many of the characteristics of colonial economies. Nevertheless, the validity of the viewpoint is still debated. Firstly, many governments of developing countries welcomed tourism as a means of stimulating economic growth. The encouragement of the growth of the industry is not typical of colonial beginnings, which are usually imposed. Secondly, most developing countries are independent politically, and foreign powers do not determine the decisions of the governments of those countries. However, the manipulation and control of local politicians and élites by foreign metropolitan interests undoubtedly exists but it is unlikely to be as influential and domineering as under colonial rule. Detailed investigations into patterns of ownership, investment, decision-making, profits, and levels of financial involvement of foreign investment would be required to substantiate either perspective. This is a daunting task but it has been started by van der Werff (1980).

The above discussion has examined some of the more important conditions which provoke intense feelings of resentment towards tourists and tourist activities. The responses of hosts to tourism also may reflect a number of other social impacts. These effects have usually been examined separately in the literature and have not been related specifically to host feelings or behaviour. These social impacts are examined below.

Tourism and moral conduct

Archer (1978: 129) believed that one of the most significant and least desirable by-products of tourism is its effects on the moral standards of the host population. The growth of prostitution, crime and gambling have been mentioned frequently as evils of tourist development.

Prostitution

Prostitution, which has been called the 'oldest profession', was as much, if not more, a part of ancient society as travel. It certainly existed prior to the growth of mass tourism. It is exceedingly difficult to say how much, if at all, tourism has been responsible for upsurges in prostitution in such tourist locations as Bali, Fiji, the Seychelles, Tonga and parts of Mexico. One might expect one or more of the following hypotheses to explain increases in prostitution in tourist resorts:

1. The processes of tourism have created locations and environments which attract prostitutes and their clients;
2. By its very nature, tourism means that people are away from the puritanical bonds of normal living, anonymity is assured away from home, and money is available to spend hedonistically. These circumstances are conducive to the survival and expansion of prostitution;
3. As tourism affords employment for women, it may upgrade their economic status. This, in turn, may lead to their liberalization and, eventually, to their involvement in prostitution to maintain or acquire new economic levels;
4. Tourism may be used as a scapegoat for a general loosening of morals.

There is, at present, little evidence to confirm or reject these assertions. Numerous studies give passing reference to the effects of tourism on prostitution (Urbanowicz 1977a,b; Archer 1978; Pizam 1978) but detailed studies do not exist. Symanski's (1974) geographical analysis of prostitution in Nevada examined the interactions between prostitutes and their clients but made no explicit reference to tourism. Information on the legal existence of brothels in the state diffuses as far as California and Utah and, undoubtedly, tourists from such distant places make use of their services. Symanski did not mention the proportion of the clientele who are tourists, or whether their patronage is necessary for the survival of these establishments.

Roebuck and McNamara (1973) were similarly inconclusive in their study of prostitution in a Mexican border city. In spite of a widespread decline in prostitution, this location exhibited a surprising occupational stability in a profession which is usually highly unstable. The authors attributed this to a number of factors which included:

1. The attractions of employment and income offered by the profession;
2. The stringent health standards which were imposed and the delimitation of areas within the city where prostitution was a legal activity;

3. The abundant clientele of American servicemen in the vicinity and the steady flow of American tourists crossing the border.

The significance of tourists when compared to the other factors contributing to this unique situation is not known.

Advertising which exploits the four Ss of tourism – sea, sun, sand and sex – through the use of erotic pictures and saucy slogans (It's better in the Bahamas!), has created images for the Caribbean and Pacific Islands as havens for sexual enjoyment. The alleged permissiveness and promiscuity of the inhabitants of these islands has even become a selling point (Turner and Ash 1975: 151). Attitudes of residents of the Seychelles towards sexuality have been called 'happily amoral'. Illegitimacy is common and partners of separated marriages frequently form new relationships without seeking divorce. These attitudes, coupled with the aspirations of local women for western economic status, form a foundation for the proliferation of prostitution. Seychelles has one of the highest rates of venereal disease in the world. The reduction of prostitution and venereal disease is not helped by the promotion of promiscuous images of this 'island of love'.

Discussions of relationships between tourism and prostitution are frequently speculative. There is a lack of evidence concerning connections between tourism, prostitution and the spread of venereal disease.

Crime

Unlike information on prostitution, data on crime are relatively easy to secure, although it is often difficult to relate the data to tourism. A wide variety of questions can be asked concerning the contribution of tourism to crime. They include:

1. Does tourism stimulate an increase in crimes against people and property? What types of crime, if any, are particularly associated with tourism?
2. Does the tourist season in resorts attract criminals from other areas and provoke the unlawful ambitions of local criminals?
3. Do people consciously consider crime when they evaluate destinations in their travel plans? Do perceived levels of crime in resorts influence travel decisions?
4. Does tourism effect the perceived levels of safety and security of inhabitants of destination areas?
5. What measures are most appropriate to ensure the safety of both hosts and guests and their property and, at the same time, maintain a carefree, holiday atmosphere?

Such questions formed the basis for discussion at a recent conference on crime detection and law stabilization in tourist regions (Nicholls, 1976). The literature on crime as an externality of tourist development is not large, but most is empirically based. McPheters and Stronge (1974), Jud (1975), Pizam (1978) and to a lesser extent Rothman (1978) indicated a positive relationship between increases in tourism and crime rates. However, with the exception of Jud, all are cautious in their interpretations because of the large number of complicating factors. Jud developed a mathematical model in his analysis of tourism and crime in Mexico but he was restricted in the variables that he was able to include. Lin and Loeb (1977: 165) indicated that because of specification errors, many of Jud's estimates could be mistaken. Nevertheless, Jud's substantial empirical evidence does seem to suggest a relationship between tourism and crime. Lin and Loeb considered the following three factors to be critical in influencing relationships between tourism and crime:

1. The population density during the tourist season;
2. The location of the resort in relation to an international border;
3. The per capita incomes of hosts and tourists, large differences between them tending to encourage robbery.

In a study undertaken in Miami, Florida, McPheters and Stronge (1974) noted a close similarity between the tourist season and the crime season. It was suggested that this reflected the response of criminals to the increased availability of targets and congestion during the tourist season, since these factors increased the potential gains and reduced the probability of detection from the point of view of the criminal. Economic crimes (robbery, larceny and burglary) had a similar season to tourism, while auto theft and crimes of passion (murder, rape and assault) did not.

Rothman's (1978) attitude studies of two seasonal resorts in Delaware indicated that inhabitants of Bethany and Rehoboth perceived a decline in personal safety and an increase in crime rates during the tourist season. It was widely believed that it was more dangerous to be out after dark during the summer than at any other time. Police protection was increased greatly during the tourist season at great expense to the local communities. Nevertheless, crime rates still increased fivefold in Rehoboth between 1974 and 1975. Pizam's (1978) study in Cape Cod also revealed that crime, particularly vandalism, drug abuse and disorderly behaviour resulting from the abuse of alcohol, was perceived as the most negative of the impacts of tourism. The development of black markets in developing countries has also been encouraged by tourism (Lehmann 1980).

In summary, it appears that tourism contributes to an increase in crime, especially on a seasonal basis. It does this through the generation of friction between the host population and tourists which may be manifest in criminal activities. In addition, the target for criminals is expanded and situations are created where gains from crime may be high and the likelihood of detection small. The effects of crime on host communities appear to be:

1. Increased expenditures on law enforcement during the tourist season;
2. Monetary losses from burglary and larceny, property damage from vandalism, commercial embezzlement, tax dodging and the growth of black markets;
3. Heightened tension;
4. The visible presence of the law in increased foot patrols and traffic controls may lead to a false sense of security.

Research priorities should be directed to the examination of the causes of increased crime in tourist areas, the indirect effects of this on the host community, the costs created and the determination of who should bear them.

Gambling

There has been a transformation of American attitudes towards gambling in the past twenty years and this has been accompanied by an unprecedented growth in the legitimate gambling industry. In 1978, 44 of the 50 American states permitted legal gambling in one form or another. The main justifications for the legalization of gambling, and its organization and promotion by the state, were the tax revenues that could be obtained, and the belief that the gambling industry, if illegal, would operate anyway. Legalization was seen as a means of diverting revenues from illegal operations to the state.

Gambling, like prostitution, is not endemic to tourism. However, it has been largely responsible for the fame and notoriety of such tourist destinations as Monte Carlo, Las Vegas and Tijuana, and for the rise and fall of Cuba's Batista in Havana. Atlantic City, New Jersey, recently voted in favour of legalizing casinos in an attempt to revive the resort's fading tourist trade (Stansfield 1978: 238). In spite of the concern on the part of local police and church groups that gambling would attract organized crime, prostitution and violence, casinos were legalized in the hope that they would:

1. Generate tourist activity;
2. Generate local employment and economic activity;
3. Increase the income of the city by expanding its tax base.

The increasing interest in the use of casino gambling as a means of reviving declining resorts has been associated with a growing concern for the problems that follow. Pizam (1978: 9) reported that, because of its side effects, gambling was perceived as one of the most undesirable consequences of tourism by residents of Cape Cod. Wall and Maccum Ali (1977: 47) reported that casinos have been banned in Trinidad and Tobago because of these problems.

The growing economic importance of gambling in the United States has seen a parallel increase in research into its many aspects (Eadington 1978: 446). Little of this research has been directly concerned with tourism but, as gambling has attracted tourists who currently support and maintain much of it through their patronage, its effects can be assumed to be partially the result of tourist activity. The contribution of gambling to local tax revenues is readily apparent. The social and psychological effects are unresolved. These include the implications of gambling for host attitudes and values, the potential created for the emergence of prostitution, crime and violence, and the extent to which gambling can grow before the market is saturated or measures are required to suppress further growth. Such questions pose a serious challenge to researchers who need to refine their research techniques to tackle the issues adequately.

Religion

Religion has been a powerful force which has long caused people to travel to religious centres in many parts of the world. Travel to the ancient cities of Palestine and Christendom, Mecca, Medina and Bangkok is not new. Pilgrimages by the Persian Shiites to their sacred death-place at Kadhimain in Iraq, by the Jews to Jerusalem and by the Sunnis to Medina, have been described by Ritter (1975: 57) as 'tourism of the dead', a form of travel which is little known in the western world. Although tourists of Western Europe do not often travel for spiritual reasons to their aspired burial place, they are often motivated to travel because of religious affiliations or curiosity. The peak tourist periods in Jerusalem and Damascus during Easter and the time of the Passover are partly a result of the spiritual devotion of western Christians. Rome or, more specifically the Vatican, attracts Catholics from all parts of the world. The cathedrals of England have become such popular tourist attractions that their very fabric is threatened by pressures from visitors (English Tourist Board 1979).

Relationships between tourism and religion have changed from their traditional form. Holy places, such as Jerusalem, Mecca and Medina, have become tourist destinations for visitors lacking a strong spiritual motivation. Anti-western sentiment has increased in such places because of political factors and because locals and devout pilgrims find that their living conditions and religious experiences have been marred because of frequent photography, the proliferation of signs and rowdy behaviour. Thus, conflict is arising between locals, the religiously devout tourist and the curious visitor. There is concern that holy places are being developed for tourism and that this is detracting from the religious significance which has made them famous.

The Church has exploited tourism for its own benefit. Tourism is being used as a source of revenue. Collection boxes are often found at entrances and exits to cathedrals and shrines, and souvenir booklets and postcards are frequently displayed for sale. In some churches guided tours are conducted and donations are requested on leaving. This situation is not the prerogative of any one religion or culture: it is as prevalent in the temples of India as in the cathedrals of Britain.

The Church has recently expressed concern over the growth of tourism because of the emergence of associated social, cultural and environmental problems. Such concerns have been expressed by the World Council of Churches (1970) and the Caribbean Ecumenical Consultation for Development (1971) in the proceedings of their meetings. They have attempted to define the role of the Church in contributing towards a more adequate understanding of tourism. They saw their role as primarily an advisory one. They proposed the following:

1. The education of the populace of host countries. This would include a reorientation of attitudes to enable hosts to fulfil their role in a positive and responsible manner;
2. Churches of host countries should initiate pastoral relationships which

would encourage visitors to share their anxieties and personal problems with a friendly and receptive host community;

3. Ensuring that tourist experiences renew the human condition, promote the perception of things not customarily perceived, and thereby promote a spiritual rethinking.

Both conferences put forward recommendations concerning the social and environmental compatibility of tourism, self renewal and spiritual awakening. Their recommendations, although ideological and perhaps unrealistic, are some of the first suggestions which attempt to tackle the social problems of the tourist industry. A third conference on 'the pastoral care of tourism on the move' (International Congress on Pastoral Care of Tourism on the Move 1979) was very wide-ranging in its deliberations, discussing such issues as the preparation of guides for holy places, the Lord's Day in the context of tourism, the pastoral reception of pilgrims, pastoral possibilities in health resorts, and the apostolate among hoteliers and young seasonal visitors. Such conferences are a clear indication that the churches have recognized that tourism is an important component of life, with both desirable and undesirable consequences worthy of their attention.

Language

The scarcity of research investigating linguistic impacts of tourism means that it is not possible to discuss this social aspect in detail. As a preface to this discussion it is appropriate to examine briefly the socio-cultural role of language in society and its relationship with tourism. Language is a vehicle of communication and is a part of the social and cultural attributes of any population. The great significance of language to society and as a socio-cultural indicator was stressed by Wagner (1958: 86) when he stated: 'Language exercises a decisive influence on the composition and distribution of inter-communicating social units – on who talks to whom – and thus on the activities in which men are able to participate in groups.' In a situation of social and cultural contact, the degree to which the native language is still spoken is an indicator of the extent of social assimilation and the strength of the culture and identity of the indigenous population. Moreover, changes in language may also be associated with changes in attitudes and behaviour on the part of local residents towards visiting groups. Language is an important factor in an analysis of social and cultural change and could be a useful indicator of the social impact of international tourism.

Only two studies were located which identified the effects of tourism on the use of language. The study by White (1974) examined relationships between the growth of tourism and social change, using language change as an index of the latter. He represented these relationships, and the possibilities of change within them, in the form of a conceptual model (Fig. 10). The model identifies three ways in which tourism can lead to language change:

1. *Through economic change.* The new jobs associated with expanding tourist

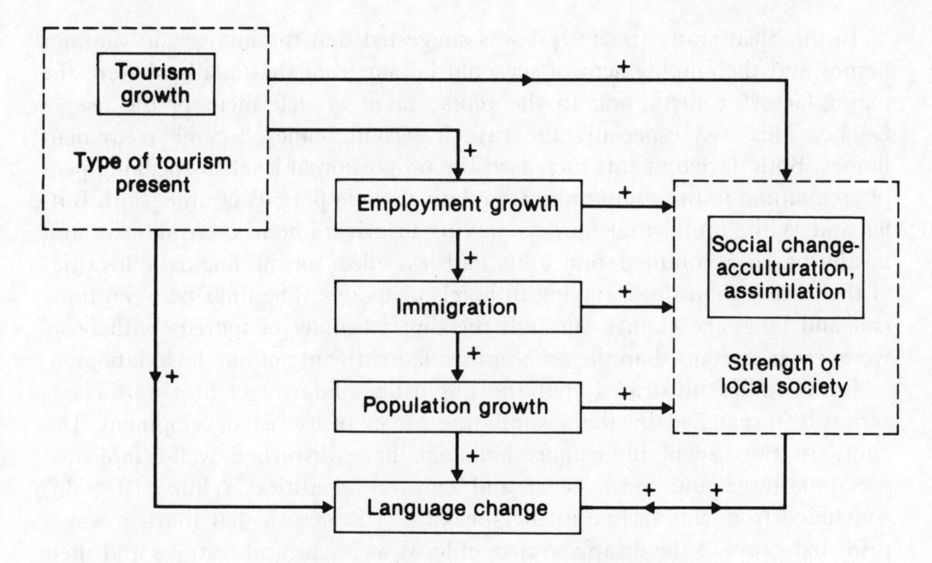

Fig. 10 The tourism–language change model (*Source*: White 1974)

development are frequently not filled by local residents and immigrants are hired. A shift towards the use of the immigrants' language may occur as they exert pressure on local residents to speak their language;

2. *Through the demonstration effect*. Tourists' portrayal of their material and financial background, and their attitudes and behaviour, may introduce new viewpoints and stimulate a broader interest in non-local affairs within the indigenous community. Aspirations of achieving similar status on the part of hosts may prompt them to replace their own language with that of the tourists;
3. *Through direct social contact*. This involves direct communication between tourists and hosts and, although it may occur under a variety of circumstances, workers in the retail and service sectors may be required to converse in the tourists' language, for the latter may not be able to speak the local tongue. White's study of twenty-eight communes in the south-east of the Swiss Canton Graubünden showed that the use of Romansh as the first language had declined markedly. For example, in areas where tourist activity was high the percentage of residents claiming Romansh as their first language declined from 66 per cent in 1888 to 29 per cent in 1970. In areas of limited tourist activity the decline was much less: from 86 per cent to 80 per cent.

Butler's (1978) study of the rural, insular, crofting parish of Sleat in the Isle of Skye, Scotland, exhibits less striking results to that of White, although both studies indicate that tourism acts to displace the indigenous language by that of the tourists. In Sleat a slight majority of residents felt that tourism had no impact upon the Gaelic language and culture. Of those who did think there was an effect, the majority felt this to be negative towards Gaelic. This was because the presence of tourists inhibited the use of Gaelic.

In the Sleat study (p. 200), it was suggested that the increase in summer homes and the employment of seasonal labour from the mainland were the major factors contributing to the replacement or deferment of the use of Gaelic. This was especially the case if second homes became retirement homes. Both developments increased the proportion of English-speaking people remaining in the community for relatively long periods of time. Both Butler and White found that tourists staying in private homes, farmhouses and locally based accommodation units had less effect on the linguistic loyalties of their hosts than those staying in hotels or motels. The links between tourism and language change through the direct contact of tourists with hosts were less important than the economic link, i.e. the use of non-local labour.

Both studies illustrated that the linguistic solidarity of host cultures is seriously threatened by the assimilating forces of tourist development. Declines in the use of indigenous language have disturbed well-established social patterns and local social and cultural identities. White (1974: 35) concluded from the decline in the speaking of Romansh that tourism was a principal cause of the disappearance of local socio-cultural features and their replacement by a greater degree of homogeneity.

Although the two studies are in substantial agreement, it may be premature to extend their findings to other areas. More detailed investigations are required for different geographical locations, for different cultural and linguistic groups, and for different types of tourist development before this can be done with assurance.

Health

Tourism and health are linked in three ways:

1. The betterment of health is a common motive for travel;
2. The standards of public health in destination areas contribute to the quality of the tourist product;
3. Diseases may be contracted by tourists in one place and transferred to other destinations or their home.

One of the most ancient links between travel and health is the taking of waters at mineral and hot springs (Lowenthal 1962: 124). Health tourism, as defined by the IUOTO (1973: 7), is 'the provision of health facilities utilizing the natural resources of the country, in particular mineral water and climate'. Many other health establishments and rest homes with minimal dependence upon natural resources, such as health farms, should be added to this definition. Health tourism originated in the belief in the curative power of climate, mineral springs and other environmental conditions. The therapeutic value of mountain air, mineral water and sunshine led to the emergence of spas in different parts of Europe. Although many of the historic spa resorts (for example Bath, Baden-Baden and Bad Gastein) have declined or developed other attractions, the importance of health tourism remains. The IUOTO, in 1972, reported the existence of 516 health resorts in Germany

(48 per cent of the European total), 290 in Poland (27 per cent), 90 in Spain (8 per cent) and 80 in Switzerland (7 per cent). The main aim of health tourism is personal regeneration through an active physical life coupled with mental relaxation. According to Wolski (1977: 203), this is an important component of the demand for tourism within contemporary society.

The second and third links between tourism and health are more pertinent to the objectives of this book. The public health conditions in the destination area have an important bearing upon the quality of the tourist product and experience. Tourists, like any other travellers crossing an international border, have a duty to conform to the health formalities in force but, at the same time, can be expected to be protected against risks during their stay. The interests of the tourists and the destination area are served by the provision of high quality health facilities which should receive the highest priority. As Richard (1973: 32) pointed out: 'People go on holiday above all for a change, to indulge in sports in a different climate and to enjoy local amenities; but no matter how excellent the amenities provided, tourists attach the utmost importance to their health and comfort.' The quality of public health facilities contributes to levels of visitor satisfaction and also determines the likelihood of tourists contracting diseases which may be transferred to other destinations or back to the place of permanent residence. For example, it is common to hear of western tourists contracting stomach ailments or dysentery from the food and water in many developing countries. This is such a common occurrence that the ailments have been given such nicknames as 'Delhi belly' and 'Montezuma's revenge'.

Tourism can also contribute to the maintenance and improvement of public health facilities in destination areas because it provides additional sources of revenue which can be invested in their upgrading. However, tourism also induces pollution which may be conducive to the development of disease. Thus, paradoxically, tourism can aid in the improvement of public health facilities in destination areas and, at the same time, lead to a deterioration in standards of health.

Summary

Social impacts of tourism were, until recently, a neglected area of study but there are signs that the topic is beginning to attract increasing attention. There has been a recent spate of studies of host attitudes to tourists, such as those undertaken by Knox (1978), Thomason *et al.* (1979) and Pearce (1980), as well as discussion based primarily on personal experience, for example, Cheng (1980). Most of these works have a poorly-developed conceptual basis and they tend to emphasize the negative social effects of tourism. However, evidence has been gathered from such diverse places as Thailand (Cohen 1979) and Colombia (Belisle and Hoy 1980) that tourism may not always be as destructive as was first thought. The issues which have been examined in this part of the book indicate the need for a more explicit account of the non-economic effects of tourism accruing to members of host communities. An examination of the effects of tourism on culture reveals similar results. These impacts are the subject of the remainder of this chapter.

Cultural impacts

It would be difficult to overlook the importance of culture as a motivator of travel. While it is easy to state the general significance of culture, it is more difficult to define the term in a way which will receive general acceptance, and, consequently, it is hard to be precise about the impacts of tourism on culture. The concept of culture has been debated in anthropological literature for at least two centuries and has acquired almost as many definitions as those trying to define it. According to Singer (1968: 540) recent definitions of culture have grown progressively more formal and abstract. Culture has often been loosely defined as behaviour as observed through social relations and material artifacts. Although these may provide some raw data for a construct of culture, they are not, in themselves, the constituents of culture. Culture, in a deeper anthropological sense, includes patterns, norms, rules and standards which find expression in behaviour, social relations and artefacts. These are the constituents of culture. Singer's definition (1968: 528) revealed this development: 'Culture consists of patterns, explicit and implicit, of and for behaviour, acquired and transmitted by symbols, constituting the distinctive achievement of human groups including their embodiments in artefacts; the essential core of culture consists of traditional (i.e. historically derived and selected) ideas and especially their attached values. . .' Thus, according to the above definition, culture is the conditioning elements of behaviour and the products of that behaviour.

Literature which examines the cultural effects of tourism concentrates upon the manifest expressions of culture from which the conditioning elements are inferred. Researchers have neglected to measure the deeper cultural effects, i.e. impacts on values, norms and identities. Graburn's (1976) commentary and collection of papers is something of an exception. It treated the material expressions of 'fourth world' peoples, which included peoples in developing areas and traditional societies in western countries, as elements of culture, reflecting changes in the culture as a whole.

This portion of the chapter examines cultural effects of tourism and includes the effects of 'cultural' tourism. The former refers to changes in the elements of culture resulting from the presence and activities of tourism. Such changes are not confined to those emanating from cultural tourism. Smith (1977: 2–3) defined cultural tourism as the absorption by tourists of features resembling the vanishing lifestyles of past societies observed through such phenomena as house styles, crafts, farming equipment and dress. Cultural tourism has been defined by Ritchie and Zins (1978: 257) as an element in the attractiveness of tourism regions. They isolated twelve elements of culture which attracted tourists to particular destinations:

1. Handicrafts;
2. Language;
3. Traditions;
4. Gastronomy;
5. Art and music, including concerts, paintings and sculpture;
6. The history of the region, including its visual reminders;
7. The types of work engaged in by residents and the technology which is used;
8. Architecture giving the area a distinctive appearance;
9. Religion, including its visible manifestations;
10. Educational systems;
11. Dress;
12. Leisure activities.

In their analysis of data gathered from people responsible for the development and operation of tourism and cultural institutions, handicrafts, gastronomy, traditions, history, architecture and leisure activities ranked consistently highest on their scale of attractiveness. Research into the cultural impacts of tourism has a similar emphasis and is centred around three major forms of culture which attract visitors:

1. Forms of culture which are inanimate or which do not directly involve human activity. Tourists visiting places of unique architecture and art, historical buildings and monuments, and purchasing traditional arts and crafts are notable examples;
2. Forms of culture reflected in the normal daily life of a destination. Visiting 'foreign' peoples to observe their normal social, economic and leisure activities in an attempt to understand their lifestyles, ideologies and customs is a common tourist motivation;
3. Forms of culture which are especially animated and may involve special events or depict historic or famous occurrences. Examples include musical festivals, carnivals, festivals reflecting old traditions and behaviour, re-enactments of battles and displays of old machinery.

Anthropological and sociological analyses of the impacts of tourism have increased rapidly in the past decade. The first anthropological symposium on tourism was held in Mexico City in 1974 and it has been conducted every year since then. Pacific tourism has been the subject of two symposia (Finney and Watson 1977; Farrell 1977). A number of texts, and academic articles in a wide range of journals, have been devoted to anthropological aspects of tourism. This can be seen as a response to the recognition in anthropological circles that tourism is a respectable subject of enquiry.

This section of the chapter concentrates upon two major themes:

1. Tourism and processes of cultural change: acculturation and cultural drift;
2. Intercultural communication and the marketing of culture, or culture expressed as a commodity.

Tourism and cultural change

Cultural change is induced both by factors which are internal and external to culture. Cultures would change in the absence of tourism. Cultural change may occur as a result of:

1. The modification of the ecological niche occupied by a society. Changes to the habitat of a society induce changes which may involve cultural adaptation and change;
2. The contact between two societies with different cultures may bring about changes to both groups;
3. The evolutionary changes occurring within a society. This refers to the process of adaptation where societies change in response to internal, biological and social needs to ensure satisfaction from their environment.

Tourism could conceivably influence all three of these modes of change but the literature on tourism documents the effects in the second category, i.e. changes induced through contact between societies with different cultures.

The tourist analyst faces severe difficulties in separating the effects of tourism on host cultures from those changes induced by other causes. Tourism represents only one form of exposure of hosts to elements of societies with a different culture. It appears that tourism is able to accelerate cultural change but the effects are not endemic to tourism and some of the modifications mentioned in the literature may reflect a series of changes, not all of which are the result of tourism.

The early work of Nunez (1963) documented the interaction between Mexican villagers and urban tourists, noting rapid and dramatic changes in economic and political structures, land use patterns, and value systems. McKean (1976: 238) described similar results in a study undertaken with Taos Pueblo Indians: 'They have come to be allied in a "cultural symbiosis" with the Anglo motel and shop operators recognizing the importance of the Indians in attracting the tourists and the Indians knowing that the whites operate necessary tourist services that enable them both to profit.'

The acculturation theoretical framework is common to these pioneering studies and to many that have been undertaken subsequently. Anthropologists have been examining acculturation for decades, arguing that when contact takes place between a strong culture and a weaker one, it is usually the former which influences the latter (Petit-Skinner 1977: 85). These studies have the underlying assumptions that cultural changes:

1. Occur primarily to the indigenous society's traditions, customs and values rather than to the visiting group;
2. Are leading to a gradual homogenization of cultures in which the local identity is being assimilated into the stronger visiting culture.

Acculturation theory asserts that when two cultures come into contact of any duration, each becomes somewhat like the other through a process of borrowing (Nunez 1977: 207). Borrowing is by no means symmetrical and is largely influenced by the nature of the contact situation, the socio-economic profiles of interacting individuals or groups, and the numerical differences in the populations. As many destinations of cultural tourism are in less developed countries, tourists, who are generally western and more wealthy, are less likely to borrow from their hosts than their hosts are from them. It seems to be inevitable that, as host societies adapt to tourism and attempt to satisfy the needs of tourists, they will succumb to tourists' attitudes and values and become more like the culture of their visitors. Most studies and examples illustrated in this section have noted a gradual weakening of traditional culture and what has been termed the development of a 'Coca-Cola' society within indigenous lifestyles (Shiviji 1973: 10).

Tourism represents one means by which acculturation can be studied. Changes resulting from intercultural contact are not exclusive to contemporary times nor are they solely attributable to recent mass tourism. Almost all communities had been exposed to outside contact before the recent upsurge in tourism. Increases in mobility, the processes of urbanization, and contact of indigenes with explorers and missionaries are but a few of the factors contributing to the breakdown of cultural barriers. It is inevitable that difficulty is encountered in separating the changes induced by tourism from those which are the result of other processes of modernization. Many souvenir art forms have been termed 'arts of acculturation' and may be the result of successive phases of modification, only the last of which is attributable to tourism.

The degree to which tourism contributes to the acculturation process, the nature and direction of effects emanating from the process, and factors which could be manipulated to minimize the role of tourism in acculturation are topics which are essentially unexplored and unanswered in the literature.

There is some ambiguity in the tourist literature concerning acculturation and its derivatives. Persistent usage has given it the meaning of cultural assimilation. Cultural assimilation means the replacement of one set of cultural traits by another (Spicer 1968: 21). The writings are inconsistent in their use of the terms which have been applied to both the effects and to the processes of change.

Cultural drift is an alternative conceptualization to explain the manifestations emanating from relationships between tourists and their hosts. It represents a relatively new approach to the study of the cultural effects of tourism. Cultural changes, as articulated through the processes of acculturation, are assumed to be the result of continuous, first-hand contacts between hosts and guests. However, relationships are frequently of a seasonal, intermittent, and multiple contact form marked by a cultural 'distance' between the hosts and guests (Collins 1978: 278). The changes that result have been called cultural drift. According to Collins:

> Cultural drift in this sense states that the role of the guest differs from that of the host and that the temporary contact situation results in change of phenotypic behaviour in both the host and the guest. The phenotypic change may be permanent in the host society/culture but temporary in the guest society/culture.

Change is initiated with the exploitation of the cultural distance between hosts and guests which, the theory suggests, still produces the normative behaviour of both groups, but with additional actions which were originally either unacceptable or constrained under previous circumstances. Contact under the cultural drift process results from both parties interacting and exploiting each other and the host environment as they strive for personal satisfaction. Phenotypic change, which is a visible change in behaviour resulting from the interaction of two groups, results. The host adjusts to the needs of the tourists when they are present but may return to a previous lifestyle on their departure. Cultural drift has the assumption that the hosts' behaviour is transformed temporarily for the duration of the relationship between hosts and guests. This is a different perspective from that of acculturation.

On the other hand, in locations where contacts between hosts and guests are more continuous or permanent, changes to the norms, values and standards of hosts may occur and these may be passed on to subsequent generations. When changes in a society or culture are handed down from one generation to the next it is known as genotypic behaviour. A situation in which both genotypic and phenotypic behaviour occur is essentially one of acculturation. If phenotypic behaviour occurs in the absence of genotypic behaviour the situation is one of cultural drift.

The adaptive process which takes place will depend upon the duration, permanence and intensity of interaction with tourists, and the strength of the host's cultural affiliations or capacity to withstand both internal and external influences of change. The debate over acculturation versus cultural drift reflects deeper and more complex conflicts in anthropology. Anthropological investigations of tourism, such as those of Graburn (1976), McKean (1977), and Smith (1977) reflect this debate. In spite of some differences of emphasis, these ideas 'open new vistas to the perception of culture change processes' (Collins 1978: 280).

Effects on culture

As in other areas of impact, there is some debate over the nature and direction of the consequences of tourism for culture, but the dominant perspective is that it is harmful. Turner and Ash (1975: 130–1) typified this perspective: 'The tourists' superior economic wealth rapidly erodes the sensuous and aesthetic wealth of cultures that have developed in isolation from the western world. . . Tourism of the present has already begun the work of obliterating cultures.' A limited number of studies stress the positive effects of tourism on culture. If support for conservation is regarded as a desirable cultural trait, then the comments of Dower (1974: 938) illustrated this viewpoint: 'Tourism and conservation can be brought to work together to mutual benefit. . . Tourism and conservation are interdependent and . . . both stand to gain from close and effective collaboration.' Thus, tourism can be viewed as a source of or remedy for problems. Both perspectives are examined below. The positive effect of tourism on culture through the promotion of intercultural communication is presented first and then the numerous other effects of tourism on culture are examined.

Intercultural communication

Mobility, which is a prerequisite of tourism, is necessary for the contact of different social groups, nationalities and cultures. It has been argued that such contacts may contribute to the removal of social or national prejudices and the promotion of better understanding and positive social change. Evans (1976: 191) postulated that cross-cultural communications between tourists and their hosts may promote adaptive changes in local culture while preserving or revitalizing local ethnic and cultural identity. Evans considered the quality of the cross-cultural communication to be of paramount importance if it is to contribute to the promotion of understanding between tourists and their hosts. She noted that the quality of cross-cultural communication could be related to a number of factors:

1. *The type of tourists*. An adaptation of Cohen's (1972) typology of tourist roles illustrated Evans's point. Institutionalized forms of tourism (organized and individual mass tourism) yield different opportunities for tourist–host interactions from non-institutionalized tourism (drifters and explorers). Variations in interaction will occur according to tourist travel arrangements and their motivations, tastes, preferences and experiences. For example, the organized mass tourist is largely confined by the itinerary of the package tour which has been purchased, and such people remain predominantly within the 'micro-environment' of their own country. On the other hand the drifter avoids developed tourist attractions and services, lives and works with members of the host culture, and shares, accepts and tries to understand their customs. In other words, non-institutionalized forms of tourism allow for more intense interaction on the part of smaller numbers of people than institutionalized forms which lead to relatively little intercultural contact;
2. *The spatial, temporal and communicative context in which contacts take place*. The nature of tourist–host interactions is influenced by the length of stay of the tourists, the time frame of actual contact, the physical and social space shared by the interacting groups, their linguistic compatibility, and the willingness of both groups to share their values, attitudes and experiences.
3. *The role of the cultural broker*. The host must also play a part in the communication process and this part may be taken by a cultural broker. The individuals in this role are usually bilingual and innovate in introducing change within their culture. In tourism they act as mediators between tourists and hosts, normally as translators and guides, and often as the organizer or retailer of goods and services which are sold for tourist consumption. Cultural brokers, therefore, have control over the amount and quality of communication between the interacting groups. They are also in a position to manipulate local culture for tourist purposes without affecting the cultural identity of the host society in a detrimental manner.

Contrary to expectations, contemporary tourism seldom generates strong intercultural relationships (Hassan 1975: 27). Many forms of tourism, particularly mass tourism, offer only incidental opportunities to learn about the social, cultural and political conditions of the destinations which are visited. Instead of destroying misconceptions between the residents of different countries, tourism perpetuates them and often creates new prejudices. Many tourists arrive with stereotypic images of their hosts and selectively perceive stimuli which will reinforce those images.

Hassan's (1975) interactional analysis of Japanese tourists in Singapore specifically examined the interactions of tourists and hosts and the factors influencing communication between them. Two out of every three tourists travelling to Singapore are members of a tour group. Hassan's study was examining, therefore, the effects of institutionalized tourism where tourists were likely to remain within their own 'environmental bubble'. Hassan's study involved tracing the movements and activities of a typical Japanese tour group and he drew the following conclusions:

1. Although many Japanese tourists desired to have meaningful contacts with local Singaporeans, they were restricted from doing so by the stringent adherence to the timetable of the package tour;
2. The duration of their stay was brief and, as a result, the Japanese preferred to hurry on to new attractions and to see as much as possible rather than to waste time talking to locals;
3. Almost all contacts which were made were highly institutionalized. Contacts were limited to shop assistants, hotel staff and other individuals servicing the requirements of the tour;
4. Language barriers intensified the problem of communication. The Japanese tourists compensated for the lack of communication by taking photographs of almost everything and everybody.

Interaction between the mass tourist and the host culture was slight, generally impersonal and superficial, and occurred primarily in the economic sphere. From the perspective of intercultural communication, Hassan (1975: 35) concluded that: 'organized mass tourism, which is the characteristic feature of Japanese tourism in Southeast Asia, does not contribute a great deal to it'. He also concluded that in mass tourism the desire to 'sightsee' restricts the opportunity for interaction between the tourists and the people who are sighted.

Information on the nature of contacts for forms of tourism other than the mass type is not readily available. More information is required concerning the influence of intercultural interactions on hosts' attitudes towards tourists, and towards tourism as a medium promoting change. Pearce (1980) found that most residents expected a high level of acceptance of foreign visitors in an area of West Virginia being considered for a tourism development programme. However, the question is an hypothetical one until those tourists actually arrive. Although intercultural communication is often commended as being an important attribute of tourism, it is clear that cultural change is a more common outcome. Miller (1974: 75) even went so far as to suggest that the tourist is a counteragent of cultural diffusion. He envisaged a gradual erosion of local culture with increases in tourist arrivals. The remaining section examines this theme.

Culture as a commodity

Most anthropological and sociological investigations have been concerned with the effects of commercialization on culture. Geographically, most studies have considered peoples in developing countries and in what Graburn (1976: 1) has called the 'Fourth World':

> The Fourth World is the collective name for all aboriginal or native peoples whose lands fall within national boundaries and technobureaucratic administrations of countries of the First, Second and Third Worlds. As such, they are peoples without countries of their own, peoples who are usually in the minority and without the power to direct the course of their collective lives.

With increasing exposure to the 'outside' world, local cultures have become 'pseudo-native', their cultures becoming more closely integrated with that of the surrounding majority.

Tourism and material forms of culture

As a result of tourism, the arts and crafts of native peoples have changed in style and form, and also in the purpose for which they are produced. Artefacts formerly produced for religious or ceremonial purposes are now produced for sale. Graburn's (1976) collection of anthropological writings on changes in Fourth World arts, emerging ethnicities, changing identities and the commercialization of cultural traditions is the most comprehensive statement on this type of tourist impact. Schadler (1979: 146–56) has written a succinct review of the topic as it pertains to Africa.

Anthropologists generally agree on the three major phases of change in traditional art forms resulting from outside contact:

1. The disappearance of traditional artistic designs and art and craft forms, particularly those with deep religious and mythical affiliations. This is followed by;
2. The growth of a degenerate, unsophisticated replacement which develops in association with mass production techniques. This is often followed by;
3. The resurgence of skilful craftsmanship and distinctive styles incorporating the deeper cultural beliefs of the host society. This phase is a response to the deleterious impacts evident in phase 2 and also to the gradual decline in the symbolic meaning of traditional arts which also occurs in the second phase.

Tourism has accelerated the promotion of pseudo-traditional arts and is largely responsible for the emergence of phase 2 of the process. However, there are also cases where tourism has induced a rejuvenation of particular forms of art and craft. The fact that many arts and crafts produced in the Fourth World are intended for external consumption indicates the formation of new relationships between peoples of the Fourth World (hosts) and art consumers of the western world (tourists). Objects which are produced in one society and are transported to and consumed in another society have been termed 'art by metamorphosis'. It is this process of transformation and its ramifications for the artefacts and their producers which form the basis of this section.

The ambivalence of tourism is expressed again in its impacts on traditional arts and crafts. Some studies have concluded that tourism contributes to the renaissance of traditional art and craft forms (Forster 1964; Hartstonge 1973; Mead 1976; Deitch 1977; McKean 1977). Others have indicated that tourism has resulted in a decline in the quality of art forms and the traditional meaning of production has been debased with production for tourist consumption (Ropponen 1976; Bascom 1976; May 1977).

The renaissance of traditional art forms

Deitch's (1977) study of the Indians of South-western United States and Graburn's (1976) work with the Eastern Canadian Inuit are clear examples of the positive effects of tourism on the arts and crafts of hosts. Arts and crafts have been a part of Indian culture in the United States for centuries. The present forms reflect an evolutionary process and a series of adaptations by host communities to new ideas, symbols and materials diffused into their culture from outside. As Deitch noted, the exposure to other ethnic groups has resulted in the widespread adoption of art traditions not endemic to the Indian. These include weaving, silverwork, shell jewellery and pottery. Following the initial borrowing, subsequent modifications and refinements by Indians have meshed together traditions of the Pueblo and Mexicans, and have led to the development of a particular form of art distinctive to the South-west Indian.

The effects of tourism in this part of the United States date back to the early 1900s, beginning with the construction of the Santa Fe railroad. Interest in the indigenous cultures and traditions of the South-west expanded with their increased exposure to Anglo-Americans, and Europeans. Tourism greatly increased the demand for Indian arts and crafts. The Indians responded with the result that there is an abundance of Indian rugs, pottery, jewellery and baskets available for purchase. Unlike in many other areas of Indian culture, arts and crafts have increased in number with continued European exposure. Their survival during early contacts, their revival with the advent of new audiences, and their manufacture using traditional technologies indicate the depth of their symbolic significance to the hosts as an important link with the past and the strength of the hosts' identity and pride in their heritage. Although there were refinements in the art forms with the infusion of new ideas, the quality and sophistication of the products has remained at the highest level.

In another study of South-west Indian art forms, particularly Navajo weaving and Pueblo pottery, Brody (1976) attributed their success to the fact that:

1. Production continues in an organic relationship with members of the tribal communities;

2. The variety of markets has meant that the arts have not been exposed to standardizing market pressures;
3. Strong positive symbolic value of the products has been maintained;
4. Trained craftsmen were available so that the effects of mechanized mass production could be avoided;
5. The culture experienced a gradual exposure to contemporary tourist pressures.

Tourism has provided employment in the fabrication of arts and crafts and induced a renaissance in the production of art forms. It has been accompanied by an improvement in the quality and artistic designs of arts and crafts. Indian art forms of the South-western United States have largely avoided the trends toward specialization, standardization and simplification which have occurred in some other host societies.

The Canadian Inuit, or Eskimo, also illustrate the rejuvenation of traditions as a consequence of tourism, but in a rather different way. Art, specifically carving, did not play an important part in Eskimo life until after the arrival of Europeans. Insatiable tourist demands for souvenirs were accompanied by an upsurge in Eskimo carving, particularly in soapstone. Graburn (1976: 42) claimed that such carvings were made solely for the profit which they generated for the host producer. Although this form of art was initiated to meet the demands of the souvenir market, it has a number of positive attributes:

1. It is of a superior quality to most souvenir art. The satisfactions gained from the occupation have superseded the initial economic motives for production;
2. The carvings draw upon the traditional, ancestral lifestyle. The figures are not imitations of western products. Subject-matters range from 'idealized self-portraits' to game animals hunted for food;
3. The carvings are a new means by which the Inuit can express the qualities of their culture which is slowly disappearing.

One aspect of South-west Indian culture resembles that of the Inuit and has contributed to what has been called a 'borrowed identity'. As economic considerations emerged as survival factors, the Indians manufactured reproductions of their sand paintings for the consumer market. Brody (1976) concluded that painting had no roots in any of the communities that produced them. Whereas the Inuit responded to consumer demands and produced crafts which reflected the innermost themes of their culture, the Indians, in the case of painting, responded solely to the market for souvenirs and developed non-characteristic forms. In both cases tourism was a stimulus for the revitalization of traditional art forms and the impetus for new creations.

The deterioration of traditional forms

Other case studies have portrayed a less positive picture of the impact of tourism on traditional art forms (Bascom 1976; May 1977). They accept that tourism has provided a market that has helped to preserve traditional art forms and keep cultures alive. Against this, tourism has encouraged the production of pseudo-traditional art forms. At their best such works can have great merit, as is the case with the prints of the Indians of the western coast of Canada. At their worst, pseudo-traditional arts, which are sometimes called 'airport art', consist of stylized works which bear only the most tenuous relationship to anything in the traditional culture. Much airport art is mass produced, often by people with little knowledge of the traditional culture who may not be members of the society whose art they purport to portray (May 1977: 125). Although May's viewpoint is an extreme one, it does represent a growing theme within the literature. The demands of tourists for cheap, exotic, portable and durable souvenirs have taken their toll on traditional art forms. Changes have occurred in the meaning of art and in its social and spiritual significance for art makers. Changes have also taken place in the size, form and function of art objects, in the methods of making them, in the materials used and in the quality of production. Bascom (1976: 306), in his analysis of African art, regretfully concluded that 'great pieces of African art are no longer produced and recent pieces are of no artistic value'.

Most researchers declaim tourist art and deplore its rapid growth mainly because of its cultural insignificance and aesthetic inadequacies when compared to traditional art forms. Four attributes of tourist art prompt this attitude and they are applicable to art forms of the Pacific, Asia and Latin America, as well as Africa:

1. Workmanship. Increased volumes of production have often been at the expense of careful, precise workmanship. The Kamba carvings of Kenya are currently manufactured in Dar-es-Salaam and bear little resemblance to those created by tribesmen. Shoddy workmanship now prevails;
2. The relationship between the art and the producer. Tourist art is usually produced in advance of a sale and differs, in this respect, from traditional African art which was individually commissioned and reflected a more personal relationship between the craftsman and the person for whom it was made. The impersonal nature of the tourist market and the growth of middlemen has removed much of the spiritual meaning from the creator's work, whether it be ivory carvings from West Africa or wooden masks of the inland tribes of East Africa;
3. Motives for art production. Art is manufactured according to the tastes of tourists. Sandelowsky (1976) recorded in personal interviews with tribesmen of the Okavango Valley in Northern Botswana that they attempt to find out what people want to purchase and then make it. Three stylistic trends have emerged, in Africa and elsewhere, which reflect the tastes of tourists:
   (a) a trend towards naturalism: many tourists like carvings of animals;
   (b) a trend towards grotesqueness. Features which tourists recognize as a component of local styles are exaggerated and distorted. According to Bascom (1977: 314), this has caused masks produced by Zambian indigenes for tourist tastes to lose much of their inspiring qualities;

(c) a trend towards gigantism. Size, regardless of quality or materials, is often a major determinant of price. Many art forms, because of the income-generating potential, are considerably larger than their predecessors and, in consequence, are not used for their traditional purposes. However, it is desirable that they are not too large to fit into a suitcase. Where the original is very large, as in the case of Indian totem poles, it may be necessary to make small replicas.

4. The quality of production, particularly the production of fakes and spurious antiquity. Misrepresentation of the age or authenticity of objects is a further effect of production for tourists. The impression that great art is characteristically old has placed age as one of the major determinants of the market value of many art forms. Faking has been an outgrowth of this attitude. Ropponen (1976: 108) reported that many Lapp goods bear the label 'handmade' and 'original', even if they are not. Loeb (1977: 185) also found fakes to be common among antiques collected by tourists in Iran. Similar findings have arisen from research undertaken in the Pacific.

Although art is not dying by any means, contemporary forms are often a degeneration of traditional styles. Brasser (1975) has argued that the disappearance of aboriginal woven basketry and the adoption and diffusion of woodsplint basketry in Indian communities of north-eastern North America is an adaption of Indian crafts to an expanded white market. Similarly, Abramson (1976: 259) stated in reference to the art forms of the native New Guineans of the Upper Sepik River:

> These artifacts exhibit elements of traditional form and iconography, but they seem to lack something. They appear to be sloppily done and the fine flowing line and rhythmic curves have been replaced by a summary, almost soulless execution. The intricate curvilinear designs and surface motifs, once carved into the wood with extreme delicacy, are now carelessly daubed on in gaudy trade-store paint. Rather than being 'primitive art', the objects are quick impressions of what art once was in this area.

Subtle changes to traditional art forms are emerging from societies in the Pacific (Maori and Sepik), Africa, and in North and South America (Inuit, Pueblo, Navaho, Cuna) who are in contact with tourism. Art objects prepared for tourists have lost much of their former meaning and the old messages they once portrayed have become simply a matter of curiosity or have been lost completely. Art in the form of pottery, sculpture or painting is an expression of the craftsman's experience, values and meaning of life and is sold as 'native symbols of identity' (MacKenzie 1977: 83). These may be symbols of the age-group of the creator, insignias of occupation, or copies of weapons of war. However, most tourist purchases are not stimulated by a genuine interest in the host culture, but are acquired as a memento of the visit and as a sign to peers of the extent of the buyer's travel experiences. This assertion supports the claim of Turner and Ash (1975: 139) that: 'even when his tourism is directly concerned with the culture and history, the tourist is not encouraged to develop any real sense of history. Nor is he encouraged to view a foreign culture as a totality (of people and environment, art and religion, past and present).'

Much of the deterioration in art forms is the result of the severance of the link between art and its traditional functions in society, be they religious or secular. This is partly a consequence of the growth of excessive demands over a short period of time, and also a reflection of non-traditional craftsmen being attracted to the industry. From an anthropological perspective, the sacrilege of religious, mythical and secular symbols and beliefs is one of the most significant impacts of tourism on native arts and crafts. Francillon's (1975: 40) study in Bali revealed that there is an increasing tendency to use religious symbols and attributes, such as pennants, bamboo poles and strips of decorated palm leaves, for the sake of tourism. Most Balinese have failed to recognize that their religion is being used by the tourist industry and that some of the most beautiful cult implements are being desecrated and transformed into decorative gimmicks.

Market forces have not only contributed to changes in art forms and their quality, they have often created a new manufacturing class which is not always of host origin. Sepik arts are no longer a specialty occupation (Abramson 1976: 256). The almost total functionalism of objects was a distinctive feature of primitive Sepik art. Art portraying religious and mythical figures was created by priests, whereas war shields were designed by warrior craftsmen and decorated by priests. Shields, which are now commercial objects, have merged into one size, suppressing both the functional qualities previously incorporated into the design and also any family, religious or mythical affiliations evident in the decoration. Aesthetic rather than functional considerations are now of paramount importance. Standardization of design has occurred to meet the requirements of visual impact. Large quantities are needed for sale and manufacturing is undertaken by most men and even women.

Tourism in New Zealand has created an occupational class which is not yet seen in other Pacific Islands. The Maori greenstone (jade) tiki or neck pendant is sought by tourists. Their manufacture is in the control of Europeans. Few contemporary Maori are able to carve the figures. The only native greenstone artefacts which can be purchased are made by students attending the Institute of Maori Arts and Crafts, which trains students in traditional carving and art forms. Until recently the manufacture of tikis was even occurring outside New Zealand, in Germany and Japan. Thus tourism has encouraged non-natives to turn to native art as a business proposition.

Conscious attempts by host societies to express their identities and to restrain the pervasive influence of tourists have contributed to the revival of old art forms. MacKenzie (1977: 84) suggested that the increase in tattooing among Samoan males exemplified this trend. New Zealand's establishment of an Institute of Maori Arts and Crafts shows that the concern over the effects of tourism on culture is being transformed into action. There is a similar school for young stone carvers at Mahabalipuram in Tamil Nadu, India. Perhaps tourist apologists are gradually being proven correct – tourism may yet bring about a revival of traditional cultures after all.

Analyses of this aspect of tourist impact are made extremely difficult by the fact that native arts have been undergoing change prior to their recent exposure to the influences of mass tourism. In fact, one would not expect a vibrant art form to remain unchanged through centuries. Furthermore, the problem of differentiating between changes in arts and crafts induced by tourism, and those induced through previous contact with outsiders hampers research immensely.

Tourism and non-material forms of culture

When tourists purchase a vacation as a package they also buy culture as a package. Regardless of how ancient or complex the destination culture, it is reduced to a few recognizable characteristics, such as arts and crafts, dance, music, buildings and special functions or ceremonies, and is promoted as a commodity (Turner and Ash 1975: 140). Marketing of specific or combinations of unique destination assets is typical of the 'product oriented' marketing approach (Wahab, Crampon and Rothfield 1976: 10). In spite of the success of this strategy, it has often conjured up inaccurate and romanticized images of destination areas and their populations. For example, New Zealand is portrayed as the home of Maori poi dances, hakas and costumes, sheep-shearing, home-spinning, knitting and gardening. These symbols are evident in travel advertisements and promotional literature on New Zealand, and contribute towards the creation of a simplistic and stereotypic image of the country. It would be an easy task to draw up a list of symbolic artifacts and customs for other countries.

The marketing of culture is at its worst in developing countries. Tourist images may be built up around illusions but they, in turn, are what tourists expect and demand when they arrive and that is what is provided. Boorstin (1961: 99) described the inevitable result: 'These "attractions" offer an elaborately contrived indirect experience, an artificial product to be consumed in the very places where the real thing is free as air... They are cultural mirages now found at tourist oases everywhere.' Traditional ceremonies, festivals and customs acquire new status and values when they are transformed into prearranged entertainment rituals. Hip-swinging 'Hula' girls greet tourists in Honolulu airport, Maori concert parties perform daily in Rotorua, and Highland bagpipers play in the background while tourists mingle in Edinburgh Castle. These have become characteristic features of the respective destinations. As Lengyel (1975: 756) summarized, the tourist sees the country or destination visited in terms of its superficially picturesque, predictably 'exotic' or 'typical' aspects, and experiences local life highly selectively and episodically. The shorter the stay, the greater the distortions of reality.

The expropriation of local culture and the exploitation of local peoples performing or portraying their culture are world-wide phenomena which are not limited to the Fourth World, although the emphases of the literature may give this impression. The commoditization of culture, be it Eskimo sculpture, Bantu shields, Spanish bullfights, Maori poi dances or peasant markets, is also evident in the tourist resorts of developed countries, where the cultural distance between hosts and guests is less marked. Greenwood's (1977: 86–107) analysis of the effects of tourism on the Alarde, the major public ritual of Fuenterrabia, Spain, is an enlightening European case which illustrates similar patterns to those of developing countries.

The Alarde is a ritual festival which commemorates Fuenterrabia's victory over the French in the siege of 1638. However, it is more than a simple commemoration. It is a re-enactment of the historic event and, until recently, involved virtually all of the town. The Alarde signifies the solidarity and unity of the village and it is one occasion on which the ideas of equality and common destiny are openly expressed. It is performed solely for those who participate. Rapid increases in Spanish tourism, and the timing of the Alarde during the peak of the tourist season, have led to a collapse of the cultural meanings of the ritual. The Alarde, traditionally a private ceremony of the town, has become a public attraction through government and commercial promotion. As the event depended upon voluntary participation, when it was undertaken for profit many of the local community no longer wished to be involved because of the erosion of the historical and ideological meanings of the festival. Although the ritual is still performed it is marred by greatly reduced levels of local participation and the resulting organizational headaches.

Although in itself a minor example, this case study is instructive. It demonstrates that the consumption of host cultures by tourists is not exclusive to societies of developing nations. It illustrates that transformation of a culture is the result of its commercial exploitation. It echoes the finding of other studies of non-material cultural manifestations in the conclusion that cultural forms lose their traditional meanings when they are modified for tourist consumption. It confirms that the commercialization of culture does not require the approval of the host society and rarely does it have the power to reverse the process. Finally, the commoditization of culture has induced other undesirable side-effects: the abandonment of traditional occupations to participate in the tourist industry; entrepreneurial competition from sophisticated, non-local, retailing organizations, and the unequal distribution of wealth. Greenwood's (1977: 107) feelings summarized the influence of tourism on the Alarde and typified the stance of much of the literature when he wrote:

> Treating culture as a natural resource or as a commodity over which tourists have rights is not simply perverse, it is a violation of the peoples' cultural rights. While some aspects of culture have wider ramifications than others, what must be remembered is that culture in its very essence is something that people believe in implicitly. By making it part of a tourist package, it is turned into an explicit and paid performance and no longer can be believed in the way it was before.

Cultural arrogance

In some destinations the demands of cultural tourism have outstripped the supply. To compensate for the lack of real cultural experiences, many destinations stage attractions so that tourists can view and experience cultural aspects of host communities. This development has become an accepted outgrowth of contemporary tourism. For example, it is not uncommon to see regular hourly concerts of native dances in Hawaii, to be able to experience a fire-walking display every evening in Fiji, or to see mock wedding ceremonies in Tunisia (Rivers 1973a: 250). The staging of contrived experiences is a way 'for the traveller to remain out of contact with foreign peoples in the very act of "sightseeing" them. They keep the natives in quarantine while the tourist in air conditioned comfort views them through a picture window' (Boorstin 1961: 99). MacCannell (1973, 1977) suggested that tourist places over-express their underlying structure and thereby disturb the sensitive expectations of the tourists. Both Boorstin and MacConnell have been critical of contrived attractions for what they offer to tourist experiences, but they make no reference to the impacts such attractions have on local culture.

The staging of cultural attractions can have both positive and negative consequences. It is positive when the staged activities divert tourists from and relieve pressures upon local people and their culture. Buck (1977: 31–2) offered a positive assessment of artificial attractions in his study of tourism and its effects on the folk culture of the Old Order Amish of Pennsylvania. He argued that the establishment of other attractions, such as the sale of souvenirs, and individuals in Amish dress participating in traditional activities, diverted tourist attention away from the real domestic and agricultural activities of the Amish. In this case staged attractions helped to preserve the culture from the pressures of tourists.

In other cases, the staging of contrived attractions has negative implications for local culture. Tourism has been accused of being 'culturally arrogant' for manipulating the traditions and customs of people to make tourist experiences more interesting and satisfying. The attractions usually only display a small and superficial part of local culture. Cultural arrogance is also displayed by tourist developers and promotors. The architectural designs of most international hotels are in western styles and, where they are found in developing countries, they often exhibit little knowledge or appreciation of the social traditions of the local inhabitants. Although some attempts have been made to relate hotel architecture to indigenous styles, they have met with only marginal success in Tunisia and East Africa. Tourists travelling under package arrangements apparently still prefer the reassurance of high-rise developments. However, the construction of simple lodgings in local styles has proved acceptable to both hosts and guests in Senegal (Saglio 1979: 321–35).

Some hotel developers have attempted to incorporate local indigenous features within western-style structures. Hotel interiors are adorned with local paintings, murals and sculpture, and symbols of present and past lifestyles. Porters, maids and tellers are often local people dressed in native costume. In Bali, where the uniqueness of the Hindu-Balinese religion is a tourist attraction, developers have copied sacred buildings for hotel decorations and have used sacred objects for furnishings (Francillon 1975: 740). Furthermore, many of the Balinese temples are used as a permanent background for stages on which sacred dances are performed for tourists entering the temple. Although these are usually honest attempts to portray local culture, as in other examples, they have abstracted local cultural qualities from their real, meaningful context.

The above examples indicate that tourism is a poor medium for intercultural communication and the preservation and rejuvenation of traditional cultural forms. Although most studies have indicated that tourism's concern with culture is generally superficial,

> It still has the potential to educate, since, in its highest form, it seeks to view and understand the origins and development of cultures. Anecdotes of tourist crassness are numerous but there are also tourists who do experience new feelings – who come to some new realization of their relation to history when visiting cultures other than their own, or observing the monuments of past cultures. A new awareness of the complexities and troubles of past or foreign cultures may conceivably turn the tourist's thoughts back to the complexities of his own culture (Turner and Ash 1975: 149).

Conclusions

The rapidly growing body of literature on the social and cultural impacts of tourism has arisen with the recognition that tourist developments do not always bring benefits to host areas. They may lead to the accentuation of existing problems and the creation of new ones. Until recently most people accepted that mass tourism was a 'good thing'; but now a lively controversy has grown about whether, in fact, indigenes are really better off for playing host to these affluent hordes (Rivers 1973b: 349).

Social impacts

If it is assumed that beneficial effects are those which are conducive to the survival of the social systems of destination areas in an unchanged form, then the social effects which have been assessed in this chapter must be considered to be predominantly negative. The effects of tourism which collectively contribute to the homogenization of societies are:

1. The overcrowding of infrastructures, accommodation, services and facilities which tourists have to share with the local population;
2. The display of prosperity amidst poverty may cause explosive situations by way of the demonstration effect;
3. The employment of non-locals in managerial and professional occupations carrying greater responsibilities and superior salaries to those occupations available to members of the host community;
4. The increase in activities deemed to be undesirable, such as prostitution, gambling and crime;
5. The gradual erosion of indigenous language and culture with increasing numbers of the host society speaking the language of their visitors.

The detrimental direction of social and cultural changes in many areas is coincident with the growth of tourism. Whether they are all attributable to

Table 17 Social impacts: future research priorities

Impact categories	Impact indicators
(i) Demographic structure	Age/sex structures, migration, resident displacement, population density and composition
(ii) Social amenities (perceptions of facilities and services)	Transportation, educational facilities, freedom of choice, preservation of lifestyle, participation versus alienation
(iii) Institutional memberships	Religious groups, social clubs, civic groups, political and sporting groups
(iv) Personal safety and privacy	Quality of police standards, crime rates, time spent with family
(v) Psychological features	Psychological stress, work satisfaction, self-expression, mobility, national or community cohesion.

tourism is another matter. It is also uncertain if the sometimes contradictory findings revealed in the case studies can be extended to other destinations, under different conditions, and with different types of tourist development. Future research must not only extend the work which has already been done, it must also seek to identify and examine social and cultural impacts which have yet to be studied. Some of these effects are listed in Table 17. A full assessment of social impact requires an appraisal of all of these impact categories.

Cultural impacts

The latter part of this chapter has documented the potential of tourism to act as an incentive for the renaissance of local culture. Whenever tourism becomes an important component of the local economy there is an increase in interest in native arts and crafts. However, it is the cultural components that have value to the tourists that have been preserved or rejuvenated and not necessarily those which are highly valued by the local culture. This type of cultural awakening has sometimes made host populations more aware of the historical and cultural continuity of their communities and this may be an enriching experience. In other cases the new appreciation of indigenous culture, the revival of ancient festivals and the restoration of cultural landmarks have emerged in ways which pose long-term threats to the existence of culture in its original form.

The section examining cultural impacts was divided into two parts. The first examined the effects of tourism in promoting intercultural communication. There appears to be only limited communication between mass tourists and their hosts. Tourism, in its present form, seldom promotes understanding between peoples of different cultures. However, little is known about the quality of communications between hosts and guests in non-institutionalized forms of tourism.

The second part of this section examined the effects of tourism on material and non-material elements of host cultures. Commercialization of culture is a consequence of tourism but changes in culture were occurring prior to the advent of tourism and continue to take place in response to other forces of modernization. Some authors claimed that the tourist art market was a positive force, that a strong symbolic value was still attached to the products, and that it was conducive to the survival of traditional culture. In other examples, the influx of tourists led to a gradual deterioration of the quality of art forms. Arts and crafts were removed from their original contexts and, in some cases, fakes have been introduced. The abrasive effects of tourism were also found in non-material art forms and produced what has been called a 'fake culture'. It has been argued that detrimental changes in art forms reflect broader changes occurring in culture and society. However, little is known of the extent to which changes in art forms can be used as an indicator of such changes.

De Kadt (1979: 14–15) has presented a concise and balanced summary of the cultural impacts of tourism:

> The frequent charge that tourism contributes to a degeneration in this field appears to be an exaggeration. Even though curio production, 'airport art', and performances of fake folklore are of course stimulated by tourist demand . . . frequently arts, crafts, and local culture have been revitalized as a direct result of tourism. A transformation of traditional forms often accompanies this development but does not necessarily lead to degeneration. To be authentic, arts and crafts must be rooted both in historical tradition and in present-day life; true authenticity cannot be achieved by conservation alone, since that leads to stultification.

Tourism has undoubtedly enabled cultures to be rehabilitated and made them known to the rest of the world. However, mass tourism has also controlled the direction in which this rejuvenation takes place and many developments may not be conducive to the survival of the cultures in which they are embedded.

This chapter has examined the effects of tourism on inhabitants of destination areas which result from their interaction with tourists. Although there are some signs of the development of a conceptual base for studies of this phenomenon, most studies have been largely descriptive. Many social and cultural impacts are difficult to quantify. Future research should be directed more explicitly at determining the perceptions and attitudes of hosts towards the presence and behaviour of tourists. A small but growing number of studies actually involve the collection of the experiences and opinions of the residents of host communities through social surveys. Unless local inhabitants are contacted, it may not be possible to identify the real significance of any change.

6 conclusions

It is inevitable that the development of tourism will induce some impacts. The very nature of tourism means that it is likely to bring about land-use conflicts and modify the economic and social characteristics of destination areas. Some of these impacts cannot be avoided completely and it is important that planners of tourism development should bear these realities in mind. This book has illuminated the considerable array of effects and problems which are the result of contemporary tourism. The continued expansion of tourism at its present rate and in its existing form is not a desirable phenomenon in most destination areas. Planning measures should be directed at restraining and redirecting growth rather than encouraging it. Of course, some newly emerging resorts will be required to relieve the pressures on existing destinations and, if they are to be successful, they must attract visitors. However, it should not be assumed that all resorts should strive to expand.

A number of factors have contributed to the undesirable situation in which many resorts find themselves:

1. Inadequate forecasting. In the desire for expansion, many overloading and congestion problems have not been foreseen or have been overlooked. It is as important to examine and, if possible, to quantify tourist pressures and capacities as it is to indicate new directions for tourist development;
2. The resource-oriented emphasis of tourist marketing. The tradition of basing tourism development upon unique and frequently fragile resources has contributed to the emergence of many overloaded destinations. Furthermore, it is common to try to stimulate demand and, when successful, this has tended to concentrate tourists into a small number of places. The spatial and temporal imbalance of tourism has contributed to the problems which are being experienced by many well-established destination areas;
3. The inefficiency of planning measures. There has been a failure to recognize the many disparate components of the tourist industry and to integrate them into an overall planning framework;
4. There has frequently been a tendency to sacrifice long-term benefits for short-term gains. This issue has been discussed in some detail by Travis 1980: 79–92);
5. Inadequate attention has been given to the appropriate scale of development. Too much emphasis has been placed upon large, prestigious projects when there is growing evidence that greater returns may accrue to

destination areas from more modest developments (Hyma and Wall 1979; Rodenburg 1980).

6. A failure to specify goals adequately. It is not always clear who the major beneficiaries of tourist development are expected to be. For example, should the objective be to satisfy the needs of tourists, to promote regional development, to decongest popular tourist destinations, or to enhance the lifestyles of residents of resorts? The strategies needed to meet each of these objectives are likely to be different.

The literature documents the urgent need to provide solutions for the numerous negative side-effects of tourist development. Enthusiasm on its own is not enough: it must be transformed into action. Planning for tourist development is a complex process which should involve a consideration of diverse economic, environmental and social structures. For example, from a social standpoint, planners should understand the complex and contradictory feelings of tourists. On the one hand, they seek change, novelty, new experiences and excitement. On the other hand, tourists frequently feel insecure and afraid in a strange environment. At the same time planners should consider the conflicting opinions of members of the host community. Many individuals involved in the tourist industry are likely to welcome tourists, whereas others may resent their presence and behaviour. From an environmental perspective, planners should recognize that some environments are more resistant to the effects of tourism than others, and that the types of tourist activity will also influence the nature of impacts. Moreover, the impacts of tourism are likely to be mixed, enhancing some components but causing deterioration in others.

Tourism, then, is a complex phenomenon which gives rise to diverse and often contradictory effects. It is easy to say that planners of tourism should maximize the benefits from tourism and minimize the costs. However, it is not possible to maximize some effects and minimize others at the same time. Trade-offs will be required and compromise will be necessary. Alternative development policies should be considered prior to the commitment of resources and assessed for their feasibility, limitations, effects and ability to satisfy the requirements of hosts and guests. De Kadt (1979: 21) indicated that he was aware of no evidence that any government has deliberately set out to assess the overall effects of alternative types of tourism in order to promote those that appear to promise the greatest net social benefits. Faced with a considerable variety of tourists and types of tourism, a country interested in tourism development needs to ask whether, and to what extent, it can match its own resources with the requirements of different types of tourists. It also needs to ask what options it faces in developing these resources. Difficulties have arisen in the establishment of measures to quantify or otherwise evaluate the qualitative aspects of tourist development. The assessment of alternative policies implies the existence of a sound knowledge base but much of the research on tourism has not been very enlightening. It is necessary that studies of tourism supply the information on which sound planning decisions can be made.

Planning for tourist impacts

The planning and marketing of tourism have been primarily oriented towards the needs of tourists and the provision of interesting and high-quality tourist experiences. Key questions of traditional planning approaches have been: how many tourists will desire or can be attracted to come to a destination and what services and facilities will they require? It is imperative that planners become less preoccupied with the visitor and devote more attention to the welfare of those being visited. Planners should be asking such questions as: how many and what type of tourists does an area want to welcome and how can tourists contribute to the enhancement of the lifestyles of residents of destination areas?

Planning for the impacts of tourism can be addressed in two ways. Firstly, planning measures can be devised to mitigate the existing impacts of tourism. Where tourism planning has been undertaken it has often been remedial, attempting to intervene after much development has taken place. As in other fields, many countries have exhibited limited ability to learn from mistakes made elsewhere, and much planning may be characterized as 'shutting the stable door after the horse has bolted' (de Kadt 1979 : 21). Most measures have been of this reactionary type. The main aim of such measures is to increase the carrying capacity of the destination area which will alleviate current tourist pressures and, at the same time, maintain or even increase the number of visitors. Some of the more common methods of manipulating the destination area include the closure of vulnerable sites and attractions, the elimination of private transportation in the core areas of resorts, the imposition of building and design regulations on new structures, the introduction of entry fees, and the establishment of barriers and paved walkways in areas which are ecologically sensitive.

Secondly, planning measures may manipulate the tourists by encouraging them to make travel decisions in certain directions. The marketing of tourism is a widespread phenomenon but the methods of marketing have received little attention by either tourism planners or researchers. The gentle manipulation of tourists could be both a positive and preventive planning approach.

The two types of measures are not mutually exclusive. If planning measures are to be directed at the decisions of tourists, knowledge of the carrying capacities of destination area sub-systems is required. For example, resentment among permanent residents over the presence of large numbers of winter vacationers in Hawaii may encourage planners to seek alternative destinations for a proportion of these tourists, thereby relieving some of the pressures in Hawaii. If alternative destinations are to be promoted, planners should be aware of the densities and conditions which prompted the Hawaiian situation and, if possible, the carrying capacities of alternative destinations in order that the Hawaiian situation is not repeated.

It is widely recognized that planning is urgently needed to alleviate the impacts of tourism. It is less apparent who should bear the planning responsibilities. The difficulty of determining who plans in specific areas or circumstances is compounded by the diversity of scales at which impacts of tourism occur. For example, should impacts occurring at a local level be the responsibility of local goverments when the effects are the outcome of tourist patronage from other parts of the country? Or should costs accruing to local areas be mitigated through finance made available from regional agencies or national governments? This debate reflects the lack of an integrated planning framework for tourist development and, in some cases, particularly in the developing world, the lack of planning organizations.

The formulation of jurisdictional frameworks for planning for the effects of tourist development is beyond the scope of this work. In fact, no one organizational structure is likely to be suitable for all eventualities in all cultures. Reference to the absence of such structures and difficulties of administration are sufficient, in themselves, to indicate the complexity of planning for tourism.

Impact research and planning

Sound research is a prerequisite of good planning. Perhaps, then, the inefficient planning of the tourist industry is partially a reflection of an inadequate research base in studies of tourism. There is certainly some truth in this deduction. Research in tourism examines what is and what has been; planning asks or states what could be or what should be. The two may not be as different as they may appear. The former provides information in order that the latter may be achieved.

There are two types of research of significance to planning (Gunn 1978b: 61):

1. *Project oriented research.* This involves the use of systematic approaches for the analysis of a specific project on a particular site. The findings have local use and value but may have little applicability to other projects or sites;
2. *Building a knowledge base.* This type of research tackles basic problems facing the tourist industry and produces results which have wider implications than the previous type of research.

This work is of the latter type and is extremely broad in its conception and content. The information contained in this investigation should be of considerable use to tourist planners. This study provides an assessment of what is and what has been. To some extent it also suggests what should not be and what could be done to remedy the undesirable consequences of what exists. It provides a knowledge base for further research and for more informed inputs into planning for tourism.

Examples of tourism research and planning

Tourism development has taken place with the guidance of and in the absence of plans. In the majority of cases detailed development plans do not exist. Where plans exist they have usually been devised by government agencies and their consultants or by private developers. Unfortunately, many of these documents are not readily accessible. In consequence it has been necessary to base this work primarily upon the academic literature. However, much valuable information is contained within planning and consulting reports and the Prince Edward Island Study (Abt Associates 1976), the Canadian Government Office of Tourism (1974) national economic study, and Butler's (1975b) work for the Inuit Tapirisat are instructive examples of such documents.

Both the costs and benefits of tourism have been identified in this text. The negative impacts of tourism, in conjunction with the emergence of environmental problems resulting from other human actions, have stimulated widespread public concern over the effects of development decisions. Futhermore, the public now demand that their concerns be incorporated into the decision-making process. This has resulted in:

1. The emergence of public participation programmes;
2. Requirements that environmental impact statements be prepared.

Together they attempt to ensure that public concerns can be voiced and that procedures are adopted which will ensure a balanced approach to the assessment of development problems.

The development and evaluation of public participation programmes in resources management are well documented (Sewell and Burton 1971; Sewell and Coppock 1977). However, with the exception of a small number of very recent isolated examples, there has been little public involvement in tourism planning. This explains the neglect of this topic in the literature on tourism. Some recent studies by governmental agencies (Canadian Government Office of Tourism 1974), consultants (for example, D'Amore Associates 1979) and academics (Pizam, 1978) have tried to assess public attitudes to specific tourist developments, but a widely accepted means of doing this has to be agreed upon. Parks Canada have recently entered into public participation and consultation programmes. Many of their planning modifications have been reactive rather than premeditated, but they are now attempting to incorporate public input into the early stages of park planning. The Riding Mountain National Park was the subject of recent public discussion and evaluation and, with some reservations (Kariel 1979; McFarlane 1979), the public participation programme for the park proved to be a successful case of citizen participation in an environmental planning project (Hoole 1978: 50). One can only applaud any attempt to solicit input from the public and to keep interested parties informed of developments, but one wonders how much can be achieved when set positions are taken with little prospect for compromise. In the case of Kouchibouguac National Park, for instance, consultation rapidly turned into confrontation.

In a theoretical sense, public participation is a positive contribution towards more effective decision-making. From a practical perspective it is difficult to arrive at decisions which are socially and enviromentally acceptable and, at the same time, economically feasible. Measurement problems compound the difficulties of reaching such decisions. A decision requires the weighting of the significance of a myriad of effects and this is difficult to do satisfactorily. Cost-benefit analysis is designed to measure the costs and benefits of projects in monetary terms and a considerable literature has developed concerning this technique (Sewell, Davis and Ross 1961). However, satisfactory procedures have yet to be developed for integrating social and environmental impacts into an economic framework. Many effects of tourism are difficult to quantify and social, economic and environmental effects are not measured in similar ways. In spite of the large literature on cost-benefit analysis, there have been few attempts to integrate the analyses of social, economic and environmental effects of tourist development to derive an overall assessment of the desirability of a project. This problem was summarized concisely by Wall and Maccum Ali (1977: 48) when they stated: ' A more definitive summary statement must await detailed studies of social and environmental impacts and the development of a means of integrating economic apples, social oranges and environmental pears.' Isard (1972) has made some progress towards devising an appropriate methodology but little has been attempted in the area of tourism impact. Butler (1975b), in a study commissioned by the Inuit Tapirisat of Canada, assessed the likely impact of a variety of types of tourist development on the northern environment and on Inuit lifestyles. This is one of the few studies which attempts to assess a wide range of possible impacts before they have occurred. Butler did not suggest a specific course of action which should followed by the Inuit. Rather, his study was designed to point out the diversity of possible types of development and their consequences so that the Inuit could adopt a development strategy which was most closely in tune with their goals.

The Prince Edward Island study

It has been indicated that there are few studies which attempt to provide a balanced assessment of tourist impact and also incorporate an assessment of the attitudes and priorities of residents of the destination area. One such study is that by Abt Associates (1976) for Prince Edward Island. It considered the major trade-offs between economic and other impacts which would be associated with alternative policy options for tourism development on the island. The study provided a comprehensive view of the impacts of tourism on the provincial economy, the island's environment and residential attitudes. A conceptual model for assessing tourism impact was devised and provided a guideline for analysis, and a number of innovative social science research methodologies and economic simulation models were applied. As in Butler's study for the Inuit, no attempt was made directly to resolve the debate over the best alternatives for tourism development, nor was an exhaustive list of possible alternatives provided. The principal concern was to suggest conclusions and recommendations which should be considered during the policy development and implementation processes.

A number of trade-off issues requiring the attention of policy-makers were outlined. These included the relative importance of tourism when compared with other sectors of the island's economy; the balance between economic benefits, social inconveniences and environmental risks; relationships between private property rights and public interests; and the costs and benefits to different regions of the island. A number of recommendations were made for consideration by the Provincial government and residents of the island. The need for a comprehensive development plan for tourism was stressed and it was indicated that this should be based upon a balance between economic, environmental and social priorities. According to the study, a careful consideration of these priorities should provide the foundation for the formulation of a policy reflecting the interests, values and aspirations of residents of Prince Edward Island.

The Prince Edward Island study is one of the few attempts to assess tourist impact from an overall perspective incorporating and integrating the results of studies from each of the economic, environmental and social impact categories. It is innovative in methodology and quite comprehensive in scope but, realistically, did not come to a single conclusion on the most appropriate form which tourist development should take. This is ultimately a political decision.

Summary

Tourism has become one of the largest and fastest-growing world industries and an economic and social phenomenon of major importance. The unprecedented expansion of tourism has given rise to a multitude of economic, environmental and social impacts which are concentrated in destination areas. These effects have become pronounced with changes in the volume and character of the tourist industry. The growth of tourism has stimulated research on many aspects of the tourist phenomenon. This volume is a synthesis of the major works concerning the economic, physical and social impacts of tourism.

Most research on tourism has been undertaken at the international and national levels. Data have been relatively easy to collect at international boundaries and this has enabled studies to be undertaken at these scales. However, tourism is not evenly distributed throughout countries, but is concentrated in localized resort areas. Generalizations derived from studies at the national scale may be misleading if applied to more restricted areas. There is a need to undertake additional investigations in destination areas where most of the impacts of tourism occur.

Tourism research has tended to emphasize economic impacts and there has been a disproportionately large number of studies examining the economic benefits of tourism. They have often overlooked the economic costs of tourism and its widespread environmental and social effects. The economic orientation of much tourism research is a reflection of the viewpoint that tourism makes a positive contribution to the economic development of destination areas. Governments, development agencies, financial organizations, planning departments, and other tourist-related bodies have supported and promoted the tourist industry in the belief that tourism will bring considerable returns on investment and aid in remedying economic problems. It is not surprising, then, to find that much research on the economic benefits of tourism is instigated, conducted or sponsored by these agencies.

The rapid development of tourism has given rise to a number of physical and social problems, and concern about the effects of tourism on host societies and their environments is growing. The adverse effects of tourism have attracted attention only recently and this is reflected in the paucity of studies of environmental and social impacts. Much of the responsibility for this neglect can be attributed to the role of promoting agencies. Tourism has been encouraged for its economic benefits with little consideration for the effects on host communities or the environments in which they are situated.

Although the body of knowledge about tourist impacts is growing, it is still highly fragmented and the findings of studies are often divergent. The tourist industry is, itself, highly fragmented. In purchasing a tourism package, the tourist is buying services from a number of different suppliers. Tourist services and facilities are also sold on an individual basis. A transportation enterprise carries the tourist to a destination, a hotel supplies accommodation and food, and a local bus company offers sightseeing tours. Although each element of the industry may remain in isolation, it has limited utility to the tourist in that form. It is the combination of these elements that constitute their value and appeal. In spite of the high level of complementarity between elements of the tourist industry, the effects of tourism on each of these elements are separate, and differ widely. The variety of components of the tourist industry causes both definitional and data-collection problems.

Research on tourism impacts has concentrated upon individual components of each of the economic, physical and social impact categories. The impacts of tourism are rarely so confined. Each component should not be considered in isolation but researchers should integrate the findings within and between each impact category. In considering specific impacts in isolation, investigations of tourism have followed separate and often divergent paths.

Research on impacts of tourism has been undertaken in many different disciplines ranging from the social sciences of economics, anthropology, geography and sociology, to the physical sciences, including botany, pedology and zoology. There has been little attempt to integrate the research themes and findings between disciplines. Each has undertaken its own research, often oblivious of the work being carried out in other disciplines. The development of interdisciplinary journals, such as *Annals of Tourism Research*, may be a sign that this sitiuation may be changing.

The highly fragmented nature of tourism makes it extremely difficult to devise a single measure of tourist impact. Different measures are used to assess different impacts. Measures of economic impact, that is monetary measures, are not appropriate to the assessment of many of the intangible social effects. Furthermore, the costs and benefits of tourism are not evenly distributed through society. The benefits to one group or individual in a community may be a cost to others in the same community. The tourist industry may expand but residents of the destination area may suffer crowding, noise, pollution and modified lifestyles. Researchers should try to identify those aspects of destination areas and their populations which have low tolerance levels to tourist activity and those which have relatively high carrying capacities.

The dynamics of impact make it difficult to assess the magnitude of change. Impacts of tourism will alter with changes in the economic and social goals of tourists and their hosts, modifications of the physical environment, and with variations in the nature of tourist activity. Spatial and temporal discontinuities between tourism and its impacts imply that effects will vary from one part of the destination to another and may occur some time after the initiation of tourism. These features have been responsible for the research emphasis on primary impacts. This has occurred at the expense of the assessment of important secondary effects.

It is usually difficult to disentangle the effects caused by tourism from those initiated by other forces of modernization. Few studies have been initiated prior to the commencement of tourist development so that base conditions have seldom been established as a yardstick against which change can be appraised. Longitudinal studies are needed which can differentiate between changes induced by tourism and those attributable to other processes.

Research on tourist impacts has produced contradictory findings. Tourist activities have both positive and negative impacts upon a destination but these may differ considerably from the effects which are occurring elsewhere. Tourism generates a diversity of experiences for tourists and their hosts but few attempts have been made to explain spatial differences in effects or to identify the key variables which give rise to these inconsistencies.

Given the obstacles to, and deficiencies of, current research, it is appropriate to identify the main themes which emerge. From the evidence which has been presented it is clear that the economic impacts of tourism are largely beneficial, the social impacts are mainly undesirable, and the environmental impacts are mixed. A growing proportion of research is empirically based but there is a lack of conceptual and methodological guidelines for the assessment of tourist impacts. The synthesis of research in this book is an attempt to provide a foundation for future studies in the hope that many of the deficiencies of existing research can be overcome and gaps in the knowledge filled.

Although our knowledge of tourist impacts is far from complete it is possible to make some general statements concerning the impacts which tourism may have. The size and level of development of a country are of particular importance, and these same characteristics are also likely to be of significance when considering destination areas within countries. De Kadt (1979: 16–17) summarized the situation as follows:

> Small countries, with relatively underdeveloped production facilities and infrastructure and relatively low levels of skills among their people, are likely to experience more negative sociocultural effects as a result of tourism development than are larger, more developed countries. Resource poor island economies with limited space are, however, precisely the ones that may have most difficulty in identifying viable development strategies which do not rely heavily on tourism.
>
> The negative sociocultural effects are likely to be reduced if the growth of tourism facilities is neither rapid nor massive, and if there is time for local populations to adjust to this activity and for tourism to fit itself to the local society. On the whole the seminar endorsed a gradualist approach, especially for small countries.

It is ironic that the destinations with the most to gain from tourism, particulary the developing countries, also appear to be the most vulnerable to its undesirable consequences. Tourism is a means of reducing employment and contributing to foreign earnings but the areas most in need of these benefits often suffer from high leakages, fragile traditional cultures and are prone to pronounced demonstration effects.

The future of tourism is paradoxical for tourists are destroying the very resources that they come to enjoy. The same is often true of destinations promoting tourism. With the rapid growth of tourism and its numerous and diverse impacts, it is imperative that planning be implemented to manage these effects. Planners should consider both the costs and benefits of tourism. Many countries and resorts promoting tourism have ignored the fact that there are limits to how much tourism a particular destination can absorb. Destinations need to consider these limits and plan their tourist industry accordingly. Research on the impacts of tourism should provide an understanding of the capacities of destination areas and of the consequences which occur both prior and subsequent to these limits being reached. Unfortunately little has yet been done to devise measures of carrying capacity for tourism.

The main objectives of tourism planning are to ensure that opportunities are available for tourists to gain enjoyable and satisfying experiences and, at the same time, to provide a means for improving the way of life of residents of destination areas. Impact research is an indispensable input to the planning of tourist destinations and can help to ensure that these goals are met.

references and bibliography

Abramson, J. A. (1976) Style change in an Upper Sepik contact situation, pp. 249–65 in Graburn, N. H. *Ethnic and Tourist Arts: Cultural Expressions from the Fourth World*. University of California Press, Berkeley.

Abt Associates (1976) *Tourism Impact Study for Prince Edward Island*. Cambridge, Mass.

Affeld, D. (1975) Social aspects of the development of tourism, pp. 109–15 in United Nations, *Planning and Development of the Tourist Industry in the ECE Region*. United Nations, New York.

Agarwal, R. K. and **Nangia, S.** (1974) *Economic and Employment Potential of Archaeological Monuments in India*. Asia Publishing House, New Delhi.

Aiken, S. R. and **Moss, M. R.** (1976) Man's impact on the natural environment of peninsular Malaysia, *Biological Conservation*, **3**, 279–82.

Airey, D. (1978) Tourism and the balance of payments, *Tourism International Research – Europe*, 3rd quarter, 2–16.

Akoglu, T. (1971) Tourism and the problem of environment, *Tourist Review*, **26**, 18–20.

Alexander, L. M. (1953) The impact of tourism on the economy of Cape Cod, Massachusetts, *Economic Geography*, **29**, 320–6.

Allen, I. (1976) Mt Kenya's fading wilderness, *Sierra Club Bulletin*, **60**(6), 5–8.

Anthony, W. (1977) Industry effects of domestic travel expenditures, workshop presentation. Travel Research Association, 8th Annual Conference, Arizona.

Apter, H. (1974) Counting the (social) cost of tourism, part 1, *The Travel Agent*, 16 Dec., 24–9.

Archer, B. H. (1972) The primary and secondary beneficiaries of tourist spending, *Tourist Review*, **27**, 42–5.

Archer, B. H. (1973) *The Impact of Domestic Tourism*, Bangor Occasional Papers in Economics No. 2. University of Wales Press, Cardiff.

Archer, B. H. (1975) *The Importance of Domestic Tourism as a Development Factor for the Developed and Developing Countries*, Tourist Research Paper No. 8. Institute of Economic Research, University College of Wales, Bangor.

Archer, B. H. (1976) Uses and abuses of multipliers, pp. 115–32 in Gearing, G. E., Swart, W. W., and Var, T. (eds), *Planning for Tourism Development: Quantitative Approaches*. Praegar, New York.

Archer, B. H. (1977a) *Tourism in the Bahamas and Bermuda: Two Case Studies*, Bangor Occasional Papers in Economics No. 10. University of Wales Press, Cardiff.

Archer, B. H. (1977b) *Tourism Multipliers: The State of the Art*, Bangor Occasional Papers in Economics No. 11. University of Wales Press, Cardiff.

Archer, B. H. (1977c) Input-output analysis: its strengths, limitations and weaknesses. Paper presented to the Travel Research Association, 8th Annual Conference, Arizona.

Archer, B. H. (1978) Domestic tourism as a development factor, *Annals of Tourism Research*, **5**, 126–41.

Archer, B. H. and **Owen, C. B.** (1972) Towards a tourist regional multiplier, *Journal of Travel Research*, **11**(2), 9–13.

Archer, B. H., Shea, S. and **Vane, R.** (1974) *Tourism in Gwynedd: An Economic Study*. Institute of Economic Research, University College of Wales, Bangor.

Bainbridge, S. (1979) *Restrictions at Stonehenge: The Reactions of Visitors to Limitations in Access: Report of a Survey*. Social Survey Division, Office of Population Censuses and Surveys, HMSO, London.

Ball, D. A. (1971) Permanent tourism: a new export diversification for less developed countries, *International Development Review*, **13**, 20–3.

Barrett, J. A. (1958) The Seaside Resort Towns of England and Wales. Unpublished PhD thesis, University of London.

Bascom, W. W. (1976) Changing African art, pp. 303–19 in Graburn, N. H. *Ethnic and Tourist Arts: Cultural Expressions from the Fourth World.* University of California Press, Berkeley.

Bayfield, N. G. (1971) Some effects of walking and skiing on vegetation at Cairngorm, pp. 469–85 in Duffey, E., and Watt, A. S. (eds), *The Scientific Management of Animal and Plant Communities for Conservation.* Blackwell, Oxford.

Bayfield, N. G. (1974) Burial of vegetation by erosion material near chairlifts in Cairngorm, *Biological Conservation*, **6**, 246–51.

Beck, B. and **Bryan, F.** (1971) This other Eden: a study of tourism in Britain, *The Economist*, **240**, 6683, supplement following p. 60.

Beed, T. W. (1961) Tahiti's recent tourist development, *Geography*, **46**, 368.

Belisle, F. J. and **Hoy, D. R.** (1980) The perceived impact of tourism by residents: a case study in Santa Marta, Colombia, *Annals of Tourism Research*, **7**, 83–101.

Bjorklund, E. M. and **Philbrick, A. K.** (1972) Spatial configurations of mental process. Unpublished paper, Department of Geography, University of Western Ontario, London, Ontario.

Bjorklund, E. M. and **Philbrick, A. K.** (1975). Spatial configurations of mental processes, pp. 57–75 in Belanger, M. and Janelle, D. G. (ed.), *Building Regions for the Future*, Notes et Documents de Recherche No. 6. Departement de Geographie, Universite Laval, Laval, Quebec, 57–75.

Bond, M. E. and **Ladman, J. R.** (1972) Tourism: a strategy for development, *Nebraska Journal of Economics and Business*, **2**, 37–52.

Bond, M. E. and **Ladman, J. R.** (1973) The tourist industry: what impact on Arizona?, *Arizona Business*, **20**(10), 20–6.

Boorstin, D. J. (1961) *The Image: A Guide to Pseudo Events in America.* Harper & Row, New York.

Boote, R. E. (1967) Coastline, pp. 127–38 in International Union for the Conservation of Nature and Natural Resources, *Ecological Impact of Recreation and Tourism upon Temperate Environments*, IUCN Proceedings and Papers, New Series No. 7. Morges, Switzerland.

Bracey, H. E. (1970) *People and the Countryside.* Routledge & Kegan Paul, London.

Brasser, T. J. (1975) *A Basketful of Indian Culture Change*, Canadian Ethnology Services paper No. 22. National Museums of Canada, Ottawa.

Britton, R. (1979) Some notes on the geography of tourism, *Canadian Geographer*, **23**, 276–82.

Britton, R. (1980) The dark side of the sun, *Focus*, **31**(2), 10–16.

Brody, J. J. (1976) The creative consumer: survival, revival and invention in Southwest Indian arts, pp. 70–84 in Graburn, N. H. *Ethnic and Tourist Arts: Cultural Expressions from the Fourth World.* University of California Press, Berkeley.

Brownrigg, M. and **Greig, M. A.** (1975) Differential multipliers for tourism, *Scottish Journal of Political Economy*, **21**, 261–75.

Brownrigg, M. and **Greig, M. A.** (1976) *Tourism and Regional Development*, Speculative Papers No. 5. Fraser of Allander Institute, Glasgow.

Bryden, J. (1973) *Tourism and Development: A Case Study of the Commonwealth Caribbean.* Cambridge University Press.

Bryden, J. and **Faber, M.** (1971) Multiplying the tourist multiplier, *Social and Economic Studies*, **20**, 61–82.

Buck, R. (1977) Making good business better: a second look at staged tourist attractions, *Journal of Travel Research*, **15**(3), 30–2.

Budowski, G. (1976) Tourism and conservation: conflict, coexistence or symbiosis, *Environmental Conservation*, **3**, 27–31.

Bugnicourt, J. (1977) Tourism with no return, *Development Forum*, **5**(5), 2–3.

Burkart, A. J. and **Medlik, S**. (1974) *Tourism: Past, Present and Future*. Heinemann, London.

Burmeister, H. (1977) Mass tourism and the environment: a closer look, *Travel Research Journal*, 21–30.

Burn, H. P. (1975) Packaging paradise, *Sierra Club Bulletin*, **60**, 25–8.

Butler, G. D. (1967) *Introduction to Community Recreation*. McGraw-Hill, New York.

Butler, R. W. (1974) Social implications of tourist development, *Annals of Tourism Research*, **2**, 100–11.

Butler, R. W. (1975a) Tourism as an agent of social change, pp. 85–90 in Helleiner, F. (ed.), *Tourism as a Factor in National and Regional Development*, Occasional Paper No. 4. Department of Geography, Trent University, Peterborough, Ontario.

Butler, R. W. (1975b) *The Development of Tourism in the North and Implications for the Inuit*, Renewable Resources Project vol. 9. Inuit Tapirisat of Canada, Ottawa.

Butler, R. W. (1978), The impact of recreation on the life styles of rural communities, *Wiener Geographische Schriften*, **51**, 187–201.

Butler, R. W. (1980), The concept of a tourist area cycle of evolution: implications for management of resources, *Canadian Geographer*, **24**, 5–12.

Canadian Government Office of Tourism (1974) *Tourism: Its Magnitude and Significance*, Research Bulletin No. 2. Ottawa.

Carbyn, L. N. (1974) Wolf population fluctuations in Jasper National Park, Alberta, Canada, *Biological Conservation*, **6**, 94–101.

Caribbean Ecumenical Consultation for Development (1971) *The Role of Tourism in Caribbean Development*. Barbados.

Carson, R. (1962) *Silent Spring*. Fawcett, Greenwich, Conn.

Charlier, R. H. (1977) Review of *Tourisme et Environnement: La Recherche d'un Equilibre*, by Arthur Haulot, *International Journal of Environmental Studies*, **10**, 141–2.

Cheng, J. R. (1980) Tourism: how much is too much? Lessons for Canmore from Banff, *Canadian Geographer*, **24**, 72–80.

Christaller, W. (1963) Some considerations of tourism location in Europe: the peripheral regions – underdeveloped countries – recreation areas, *Papers of the Regional Science Association*, **12**, 95–105.

Clare, P. (1971) *The Struggle for the Great Barrier Reef*. Collins, London.

Clawson, M. and **Knetsch, J**. (1966) *Economics of Outdoor Recreation*. Johns Hopkins University Press, Baltimore.

Clement, H. G. (1961) *The Future of Tourism in the Pacific and the Far East*. US Department of Commerce, Washington, DC.

Clement, H. G. (1967) The impact of tourist expenditures, *Development Digest*, **5**, 70–81.

Clout, H. D. (1971) Second homes in the Auvergne, *Geographical Review*, **61**, 530–53.

Cohen, E. (1971) Arab boys and tourist girls in a mixed Jewish Arab community, *International Journal of Comparative Sociology*, **12**, 217–33.

Cohen, E. (1972) Toward a sociology of international tourism, *Social Research*, **39**, 164–82.

Cohen, E. (1974) Who is a tourist? A conceptual clarification, *Sociological Review*, **22**, 527–53.

Cohen, E. (1978) The impact of tourism on the physical environment, *Annals of Tourism Research*, **5**, 215–37.

Cohen, E. (1979) The impact of tourism on the hill tribes of northern Thailand, *Internationales Asienforum*, **10**, 5–38.

Collins, L. R. (1978) Review of *Hosts and Guests: An Anthropology of Tourism*, *Annals of Tourism Research*, **5**, 278–80.

Coppock, J. T. (1977) *Second Homes: Curse or Blessing*. Pergamon, New York.

Countryside Commission (1974) *Farm Recreation and Tourism in England and Wales*. DART Publication No. 14.

Crittendon, A. (1975) Tourism's terrible toll, *International Wildlife*, **5**(3), 4–12.
Cullington, J. (1980) The Public Use of Private Land for Recreation. Unpublished MA thesis, University of Waterloo, Ontario.
Cunningham, H. (1980) *Leisure in the Industrial Revolution*. Croom Helm, London.
Curwen, M. (1973) Botswana's best bet, *African Development*, 7 Feb.
D'Amore, L. J. (1976) The significance of tourism in Canada, *Business Quarterly*, **41** (3), 27–35.
D'Amore, L. J. and Associates Ltd (1979) *Social Dimensions of Environmental Planning-An Annotated Bibliography*, Report No. 17. Office of the Science Advisor, Environment Canada, Ottawa.
Darling, F. F. and **Eichhorn, N. D.** (1967) The ecological implications of tourism in national parks, pp. 98–101 in International Union for the Conservation of Nature and Natural Resources, *Ecological Impact of Recreation and Tourism upon Temperate Environments*, IUCN Proceedings and Papers, New Series No. 7. Morges, Switzerland.
De Grazia, S. (1964) *Of Time, Work and Leisure*. Doubleday Anchor, New York.
Deitch, L. I. (1977) The impact of tourism upon the arts and crafts of the Indians of the Southwestern United States, pp. 173–84 in Smith, V. *Hosts and Guests: An Anthropology of Tourism*. University of Pennsylvania Press, Philadelphia.
De Kadt, E. (1979) *Tourism – Passport to Development?* Oxford University Press, New York.
Demars, S. (1979) British contributions to American seaside resorts, *Annals of Tourism Research*, **6**, 285–93.
Diamond, J. (1976) Tourism and development policy: a quantitative appraisal, *Bulletin of Economic Research*, **28**, 36–50.
Diamond, J. (1977) Tourism's role in economic development: a case reexamined, *Economic and Cultural Change*, **25**, 539–53.
Diem, A. (1980) Valley renaissance in the high Alps, *Geographical Magazine*, **52**, 492–7
Ditmars, E. E. (1973) The fuel crisis and air travel, *Proceedings of the Travel Research Association*. 5th Annual Conference, Williamsburg, Virginia, 17–20.
Dower, M. (1973) Recreation, tourism and the farmer, *Journal of Agricultural Economics*, **24**, 465–77.
Dower, M. (1974) Tourism and conservation: working together, *Architects Journal*, **18**, 159, 938–63.
Doxey, G. V. (1975) A causation theory of visitor-resident irritants: methodology and research inferences, *Proceedings of the Travel Research Association*, 6th Annual Conference, San Diego, California, 195–8.
Doxey, G. V. (1976) When enough's enough: the natives are restless in Old Niagara, *Heritage Canada*, **2**(2), 26–7.
Driver, B. L. and **S. R. Tocher** (1974) Toward a behavioral interpretation of recreational engagements, with implications for planning, pp. 9–31 in Driver, B. L. (ed.), *Elements of Outdoor Recreation Planning*. University of Michigan Press, Ann Arbor.
Eadington, W. (1978) Gambling and society: interdisciplinary studies on the subject of gambling, *Annals of Tourism Research*, **5**, 444–9.
Economist Intelligence Unit (1973) The role of tourism in economic development: is it a benefit or burden?, *International Tourism Quarterly*, No. 2, 53–68.
Edgell, D. L. (1977) International business prospects for the rest of the century: international tourism and travel, *Travel Research Journal*, 33–9.
Ehemann, J. (1977) What kind of place is Ireland, *Travel Research Journal*, **16**, 1977, 28–30.
Ehrlich, P. A. (1970) *Population, Resources and Environment*. Freeman, San Francisco.
Elkan, W. (1975) The relation between tourism and employment in Kenya and Tanzania, *Journal of Development Studies*, **11**, 123–30.
English Tourist Board (1979) *English Cathedrals and Tourism: Problems and Opportunities*. London.

Erbes, R. (1973) *International Tourism and the Economy of Developing Countries.* Organisation for Economic Cooperation and Development, Paris.

Evans, N. (1976) Tourism and cross cultural communication, *Annals of Tourism Research*, **3**, 189–98.

Farrell, B. H. (ed.) (1977) *The Social and Economic Impact of Tourism on Pacific Communities.* Center for South Pacific Studies, University of California, Santa Cruz.

Finney, B. R. and **Watson, A.** (eds) (1977) *A New Kind of Sugar: Tourism in the Pacific.* East-West Technology and Development Institute, East-West Center, Honolulu.

Forster, J. (1964) The sociological consequences of tourism, *International Journal of Comparative Sociology*, **5**, 217–27.

Fox, M. (1977) The social impact of tourism: a challenge to researchers and planners, pp. 27–48 in Finney, B. R. and Watson, A. *A New Kind of Sugar: Tourism in the Pacific.* Center for South Pacific Studies, University of California, Santa Cruz.

Francillon, G. (1975) Tourism in Bali – its economic and socio-cultural impact: three points of view, *International Social Science Journal*, **27**, 723–52.

Frechtling, D. C. (1975) Measuring travel-generated employment, *Proceedings of the Travel Research Association*, 6th Annual Conference, San Diego, California, 43–6.

Frechtling, D. C. (1976) Proposed standard definitions and classifications for travel research, *Proceedings of the Travel Research Association*, 7th Annual Conference, Boca Raton, Florida, 59–68.

Friedl, J. (1972) Changing economic emphasis in an Alpine village, *Anthropological Quarterly*, **45**, 147–57.

Ghali, M. A. (1976) Tourism and economic growth: an empirical study, *Economic Development and Cultural Change*, **24**, 527–38.

Gilbert, E. W. (1939) The growth of inland and seaside health resorts in England, *Scottish Geographical Magazine*, **55**, 16–35.

Gilbert, E. W. (1949) The growth of Brighton, *Geographical Journal*, **114**, 30–52.

Gilbert, E. W. (1954) *Brighton: Old Ocean's Bauble.* Methuen, London.

Gillmor, D. A. (1973) Irish holidays abroad: the growth and destinations of chartered inclusive tours, *Irish Geography*, **6**, 618–25.

Goffe, P. (1975) Development potential of international tourism, how developing nations view tourism, *Cornell Hotel and Restaurant Administration Quarterly*, **16**, 24–31.

Goldsmith, E. (1974) Pollution by tourism, *The Ecologist*, **48**(1), 47–8.

Graburn, N. H. (ed.) (1976) *Ethnic and Tourist Arts: Cultural Expressions from the Fourth World.* University of California Press, Berkeley.

Gray, F. (1973) Review of *The Long African Day* by Norman Myers, *New York Review of Books*, **20**(4), 25–9.

Gray, H. P. (1970) *International Travel: International Trade.* Heath Lexington Books, Lexington, Mass.

Gray, H. P. (1974) Towards an economic analysis of tourism policy, *Social and Economic Studies*, **23**, 386–97.

Greenwood, D. (1972) Tourism as an agent of change: a Spanish Basque case study, *Ethnology*, **11**, 80–91.

Greenwood, D. (1976) Tourism as an agent of change, *Annals of Tourism Research*, **3**, 128–42.

Greenwood, D. (1977) Culture by the pound: an anthropological perspective on tourism as cultural commoditization, pp. 86–107 in Smith, V. *Hosts and Guests: An Anthropology of Tourism.* University of Pennsylvania Press, Philadelphia.

Gunn, C. A. (1978a) Needed: an international alliance for tourism, recreation, conservation, *Travel Research Journal*, **2**, 3–9.

Gunn, C. A. (1978b) Research applied to planning, *Proceedings of the Travel Research Association*, 9th Annual Conference, Ottawa, 61–3.

Haines, G. H. (1976) The problem of the tourist, *Housing and Planning Review*, **32**, 7–11.

Hall, J. (1974) The capacity to absorb tourists, *Built Environment*, **3**, 392–7.

Hall, P. (1970) A horizon of hotels, *New Society*, **15**, 389, 445.

Harmston, I. K. (1969) The importance of 1967 tourism to Missouri, *Business and Government Review*, **10**, 5–13.

Harrington, L. (1971) The trouble with tourism unlimited, *New Statesman*, **82**, 2106, 176.

Hartstonge, J. E. (1973) Arts and crafts, *Proceedings of the Pacific Area Travel Association Workshop*, Kyoto, Japan, 217–19.

Hassan, R. (1975) International tourism and intercultural communication, *Southeast Asian Journal of Social Sciences*, **3**(2), 25–37.

Henderson, D. M. (1975) *The Economic Impact of Tourism: A Case Study in Greater Tayside*, Research Report No. 13. Tourism and Recreation Research Unit, University of Edinburgh.

Hoole, A. (1978) Public participation in park planning: the Riding Mountain case, *Canadian Geographer*, **22**, 41–50.

Houghton-Evans, W. and **Miles, J. C**. (1970) Environmental capacity in rural recreation areas, *Journal of the Town Planning Institute*, **56**, 423–7.

Hunt, J. D. (1975) Image as a factor in tourism development, *Journal of Travel Research*, **13**(3), 1–7.

Hutchinson, A. (1975) Elephant survival: two schools of thought, *Wildlife*, **17**, 104–7.

Hutchinson, J. (1980) Analysis of changing tourist accommodation distribution in Toronto using point pattern techniques. Unpublished MA thesis, University of Waterloo, Ontario.

Huxley, J. (1961) *The Conservation of Wildlife and Natural Habitats in Central and East Africa*. UNESCO, Paris.

Hyma, B. and **Wall, G**. (1979) Tourism in a developing area: the case of Tamil Nadu, India, *Annals of Tourism Research*, **6**, 338–50.

International Congress on Pastoral Care of Tourism on the Move (1979) Proceedings, *On the Move*, **24**, 1–113.

Ironside, G. (1971) Agricultural and recreational land use in Canada: potential for conflict or benefit, *Canadian Journal of Agricultural Economics*, **19**, 1–12.

Isard, W. (1972) *Ecological-economic Analysis For Regional Developments: some initial explorations with particular reference to recreational-resource use and environmental planning*. Free Press, New York, 1972.

IUOTO (1963) *Conference on International Travel and Tourism*. United Nations, Geneva.

IUOTO (1973) *Health Tourism*. United Nations, Geneva.

Jackson, R. T. (1973) Problems of tourist industry development on the Kenyan coast, *Geography*, **58**, 62–5.

Jafari, J. (1974) The socio-economic costs of tourism to developing countries, *Annals of Tourism Research*, **1**, 227–59.

Jensen, G. (1979) Its called wearing out your welcome, *The Sunday Sun*, Toronto, 25 February, 38.

Jones, S. B. (1933) Mining and tourist towns in the Canadian Rockies, *Economic Geography*, **9**, 368–78.

Jonish, J. E. and **Peterson, R. E**. (1973) Impact of tourism: Hawaii, *Cornell Hotel and Restaurant Administration Quarterly*, **14**, 5–12.

Jordan, J. W. (1980) The summer people and the natives: some effects of tourism in a Vermont vacation village, *Annals of Tourism Research*, **7**, 34–55.

Jud, G. D. (1974) Tourism and economic growth in Mexico since 1950, *Inter-American Economic Affairs*, **28**, 19–43.

Jud, G. D. (1975) Tourism and crime in Mexico, *Social Science Quarterly*, **56**, 324–30.

Kariel, H. G. (1979) Public participation in National Park planning, Comment No. 1, *Canadian Geographer*, **23**, 172–3.

Kaspar, C. (1977) Social needs and their realization in tourism, *Proceedings of the Association Internationale d'Experts Scientifiques du Tourisme* (Berne, Switzerland), **18**, 19–20.

Kitchener-Waterloo Record (March 1979) Tourists are threatening wildlife park.

Knox, J. (1978) *Classification of Hawaii Residents' Attitudes Toward Tourists and Tourism*, Tourism Research Project Occasional Paper No. 1. University of Hawaii at Manoa, Honolulu.

Koea, A. (1977) Polynesian migration to New Zealand, pp. 68–9 in Finney, B. R. and Watson, A. *A New Kind of Sugar: Tourism in the Pacific*. Center for South Pacific Studies, University of California, Santa Cruz.

Konrad, V. (in press) Historical artifacts as recreational resources, in Wall, G., and Marsh, J., *Recreational Land Use: A Canadian History*. Carleton Library Series, Carleton University, Ottawa.

Kraus, O. (1967) Ecological impact of tourism and recreation on the countryside adjacent to highways, pp. 161–9 in International Union for the Conservation of Nature and Natural Resources. *Ecological Impact of Recreation and Tourism upon Temperate Environments*, IUCN Proceedings and Papers, New Series No. 7. Morges, Switzerland.

Kreutzwiser, R. D. (1978) Socio-economic impact of Walt Disney World, Central Florida. Unpublished paper, Department of Geography, University of Guelph, Ontario.

Krishnaswamy, J. (1978) The economic impact of tourism – a case study of Maharashtra State, India, *Travel Research Journal*, **2**, 17–22.

LaPage, W. F. (1974) Campground trampling and ground cover response, pp. 237–45 in Fischer, D. W., Lewis, J., and Priddle, G. (eds), *Land and Leisure*. Maaroufa Press, Chicago.

Lavery, P. (1971) Resorts and recreation, pp. 167–95 in Lavery, P. (ed.), *Recreational Geography*. David and Charles, Newton Abbot.

Lawson, F. and **Baud-Bovy M**. (1977) *Tourism and Recreational Development*. Architectural Press, London.

Lehmann, A. C. (1980) Tourists, black markets and regional development in West Africa, *Annals of Tourism Research*, **7**, 102–19.

Leiper, N. (1979) The framework of tourism: towards a definition of tourism, tourist and the tourist industry, *Annals of Tourism Research*, **6**, 390–407.

Lengyel, P. (1975) Tourism in Bali – its economic and socio-cultural impacts. A rejoinder, *International Social Science Journal*, **27**, 753–7.

Lerner, S. C. (1977) Social impact assessment: some hard questions and basic techniques. Unpublished workshop paper, University of Waterloo, Ontario.

Levitt, K. and **Gulati I.** (1970) Income effect of spending: mystification multiplied: a critical comment on the Zinder Report, *Social and Economic Studies*, **19**, 326–43.

Lewis, G. (1972) *The Virgin Islands*. Northwestern University Press, Evanston, Ill.

Lickorish, L. J. and **Kershaw, A. G**. (1958) *The Travel Trade*. Practical Press, London.

Lin, V. L. and **Loeb, P. D**. (1977) Tourism and crime in Mexico: some comments, *Social Science Quarterly*, **58**, 164–7.

Little Inc., A. D. (1962) *Tourism and Recreation*. US Department of Commerce, Washington, DC.

Loeb, L. D. (1977) Creating antiques for fun and profit: encounters between Iranian Jewish merchants and touring coreligionists, pp. 185–92 in Smith, V. *Hosts and Guests: An Anthropology of Tourism*. University of Pennsylvania Press, Philadelphia.

Lowenthal, D. (1962) Tourists and thermalists, *Geographical Review*, **52**, 124–7.

Lundberg, D. E. (1971) Why tourists travel, *Cornell Hotel and Restaurant Administration Quarterly*, **11**(4), 75–81.

Lundberg, D. E. (1972) *The Tourist Business*. Cahners, Boston.

Lundberg, D. E. (1974) Caribbean tourism: social and racial tensions, *Cornell Hotel and Restaurant Administration Quarterly*, **15**(1), 82–7.

Lundgren, J. O. J. (1973) Tourist impact/island entrepreneurship in the Caribbean. Paper presented to the Conference of Latin American Geographers.

McCloy, D. B. (1975) Employment research in the Canadian travel industry,

Proceedings of the Travel Research Association, 6th Annual Conference, San Diego, California, 49–51.

MacCannell, D. (1973) Staged authenticity: arrangement of social space in tourist settings, *American Journal of Sociology*, **79**, 586–603.

MacCannell, D. (1977) Tourist and the new community, *Annals of Tourism Research*, **4**, 208–15.

McCool, S. F. (1978) Recreation use limits: issues for the tourism industry, *Journal of Travel Research*, **17**(2), 2–7.

McEachern, J. and **Towle, W. L**. (1974) *Ecological Guidelines for Island Development*. IUCN, Morges, Switzerland.

McFarlane, R. N. (1979) Public participation in National Park planning, Comment No. 1, *Canadian Geographer*, **23**, 173–6.

McIntosh, R. W. (1977) *Tourism: Principles, Practices and Philosophies*. Grid Inc., Columbus, Ohio, 2nd ed.

McKean, P. F. (1976) Tourism, culture change and culture conservation in Bali, pp. 237–45 in Banks, D. J. (ed.), *Changing identities in Modern S.E. Asia and World Anthropology*. Mouton, The Hague.

McKean, P. F. (1977) From purity to pollution? A symbolic form in transition: the Balinese Ketjak, pp. 293–302 in Becker, A., and Yengoyen, A. (eds), *The World Imagination of Reality: Symbol Systems in Southeast Asia*. University of Arizona Press, Tucson.

MacKenzie, M. (1977) The deviant art of tourism: airport art, pp. 83–5 in Farrell, B. H. *Social and Economic Impact of Tourism on Pacific Communities*. Center for South Pacific Studies, University of California, Santa Cruz.

McPheters, L. R. and **Stronge, W. B**. (1974) Crime as an environmental externality of tourism: Florida, *Land Economics*, **50**, 288–92.

Manning, F. E. (1979) Tourism and Bermuda's black clubs: a case of cultural revitalization, pp. 157–76 in De Kadt, E. *Tourism – Passport to Development?* Oxford University Press, New York.

Marsh, J. S. (1975a) Hawaiian tourism: costs, benefits, alternatives, *Alternatives*, **4**(3), 34–9.

Marsh, J. S. (1975b) Tourism and development: the East African case, *Alternatives*, **5**(1), 15–22.

Martinka, C. (1974) Preserving the natural state of grizzlies in Glacier National Park, *Wildlife Society Bulletin*, **2**, 13–17.

Mather, A., Ritchie, W. and **Crofts, R**. (1973) Highland sand lands, *Geographical Magazine*, **45**, 863–7.

Matthews, H. G. (1977) Radicals and third world tourism: a Caribbean case, *Annals of Tourism Research*, **5**, 20–9.

May, R. J. (1977) Tourism and the artefact in Papua New Guinea, pp. 125–33 in Finney, B. R. and Watson, A. *A New Kind of Sugar: Tourism in the Pacific*. Center for South Pacific Studies, University of California, Santa Cruz.

Mead, S. M. (1976) The production of native art and craft objects in contemporary New Zealand society, pp. 285–98 in Graburn, N. H. *Ethnic and Tourist Arts: Cultural Expressions from the Fourth World*. University of California Press, Berkeley.

Mercer, K. C. R. (1976a), Why do people take holidays, *New Society*, **37**, 724, 438–40.

Mercer, K. C. R. (1976b), The application of motivational research to tourism, *Tourist Review*, **31**(4), 10–11.

Mercer, K. C. R. (1977) Needs, motives, recreation and tourism, *Rural Recreation and Tourism Abstracts*, **2**, 1–5.

Middleton, M. (1971) New uses of old towns – conservation in action, pp. 37–40 in British Tourist Authority, *Tourism and Environment*. London.

Miller, J. J. B. (1974) The tourist as the counter-agent in cultural diffusion, pp. 75–81 in Evenden, L. J. and Cunningham, F. F. (eds), *Cultural Discord in the*

Modern World, British Columbia Geography Series No. 20. Simon Fraser University, Vancouver.

Mings, R. C. (1969) Tourism's potential for contributing to the economic development in the Caribbean, *Journal of Geography*, **68**, 173–7.

Minnesota Department of Economic Development (1977) *Economic Impact of the Minnesota Tourist and Travel Industry 1976*, Research Paper No. 36. St Paul, Minnesota.

Mitchell, B. (1979) *Geography and Resource Analysis*. Longman, London.

Mitchell, F. (1970) The value of tourism in East Africa, *East African Economic Review*, **2**, 1–21.

Moment, G. B. (1969) Bears: the need for a new sanity in wildlife management, *Bio-Science*, **18**, 1105–8.

Moore, K. (1970) Modernization in a Canary Island village: an indicator of social change in Spain, *Journal of the Steward Anthropological Society*, **2**, 19–34.

Mountfort, G. (1974) The need for partnership: tourism and conservation, *Development Forum*, April, 6–7.

Mountfort, G. (1975) Tourism and conservation, *Wildlife*, **17**, 30–3.

Murphy, P. E. (1975) The role of attitude in the choice of recreation boaters, *Journal of Leisure Research*, **7**, 216–24.

Myers, N. (1972) National parks in savannah Africa, *Science*, **178**, 4067, 1255–63.

Myers, N. (1973a) The people crunch comes to Africa, *Natural History*, **82**, 10–15, 79–82.

Myers, N. (1973b) Impending crisis for Tanzanian wildlife, *National Parks and Conservation Magazine*, **47**(8), 18–23.

Myers, N. (1975) The silent savannahs, *International Wildlife*, **5**(5), 5–10.

Myers, N. (1976) Whither African wildlife?, *Sierra Club Bulletin*, **60**(6), 4–12.

Netboy, A. (1975) Tourism and wildlife conservation in East Africa, *American Forests*, **81**(8), 25–7.

Nettekoven, L. (1979) Mechanisms of intercultural interaction, pp. 135–45 in De Kadt, E. *Tourism – Passport to Development?* Oxford University Press, New York.

Neulinger, J. and **Breit, M**. (1971) Attitude dimensions of leisure: a replication study, *Journal of Leisure Research*, **3**, 108–15.

New Brunswick Department of Tourism (1974) *Employment in the Tourist Industry: New Brunswick 1972*. Technical Services Division, St John.

Newcomb, R. M. (1979) *Planning the Past: Historical Landscape Resources and Recreation*. Shoe String Press, Hamden, Conn.

Nicholls, L. L. (1976) Tourism and crime, *Annals of Tourism Research*, **3**, 176–81.

Nicholson, M. (1970) *The Environmental Revolution*. Hodder & Stoughton, London.

Noake, D. W. (1967) Camping as a factor in the ecological impact of tourism and recreation, pp. 225–9 in International Union for the Conservation of Nature and Natural Resources, *Ecological Impact of Recreation and Tourism upon Temperate Environments*, IUCN Proceedings and Papers, New Series No. 7. Morges, Switzerland.

Nolan, S. D. (1976) Tourists' use and evaluation of travel information sources: summary and conclusions, *Journal of Travel Research*, **14**(3), 6–8.

Noronha, R. (1976) *Review of the Sociological Literature on Tourism*. World Bank, New York.

Nunez, T. A. (1963) Tourism, tradition, and acculturation: weekendismo in a Mexican village, *Ethnology*, **2**, 347–52.

Nunez, T. A. (1977) Touristic studies in anthropological perspective, pp. 207–16 in Smith, V. *Hosts and Guests: An Anthropology of Tourism*. University of Pennsylvania Press, Philadelphia.

Ogilvie, F. W. (1933) *The Tourist Movement*. P. S. King, London.

Organisation for Economic Co-operation and Development (1980) *The Impact of Tourism on the Environment*. Paris.

Ouma, J. P. B. (1970) *Evolution of Tourism in East Africa*. East African Literature Bureau, Nairobi, 1970.

Ovington, J. D., Groves, K. W., Stevens, P. R. and **Tanton, M. T**. (1972) *A Study of the Impact of Tourism at Ayers Rock–Mt Olga National Park*. Department of the Interior, Canberra.

Owen, J. S. (1969) Development and consolidation of Tanzanian National Parks, *Biological Conservation*, **1**, 156–8.

Pacione, M. (1977) Tourism: its effects on the traditional landscape in Ibiza and Formentera, *Geography*, **62**, 43–7.

Parks Canada (1970) *The Economic Impact of National Parks in Canada*. Department of Indian and Northern Affairs, Ottawa.

Parsons, J. J. (1973) Southward to the sun: the impact of mass tourism on the coast of Spain, *Yearbook of the Association of Pacific Coast Geographers*, **35**, 129–46.

Patmore, J. A. (1968) The spa towns of Britain, pp. 47–69 in Beckinsale, R. P. and Houston, H. M. (eds), *Urbanization and its Problems*. Blackwell, Oxford.

Pearce, D. G. (1978) Form and function in French resorts, *Annals of Tourism Research*, **5**, 142–56.

Pearce, D. G. (1979) Towards a geography of tourism, *Annals of Tourism Research*, **6**, 245–72.

Pearce, J. A. (1980) Host community acceptance of foreign tourists: strategic considerations, *Annals of Tourism Research*, **7**, 224–33.

Perez, L. A. (1974) Aspects of underdevelopment: tourism in the West Indies, *Science and Society*, **37**, 473–80.

Perez, L. A. (1975) Tourism in the West Indies, *Journal of Communications*, **25**, 136–43.

Peters, M. (1969) *International Tourism*. Hutchinson, London.

Petit-Skinner, S. (1977) Tourism and acculturation in Tahiti, pp. 85–7 in Farrell, B. *Social and Economic Impact of Tourism on Pacific Communities*. Center for South Pacific Studies, University of California, Santa Cruz.

Pigram, J. J. (1977) Beach resort morphology, *Habitat International*, **2**, 525–41.

Pizam, A, (1978) Tourism's impacts: the social costs to the destination as perceived by its residents, *Journal of Travel Research*, **16**(4), 8–12.

Plog, S. C. (1977) Why destination areas rise and fall in popularity, pp. 26–8 in Kelly, E. M. (ed.). *Domestic and International Tourism*, Institute of Certified Travel Agents, Wellesley, Mass.

Plummer, D. (1971) London and tourism: problems and opportunities, pp. 30–3 in British Tourist Authority, *Tourism and Environment*. London.

Pollock, N. C. (1971) Serengeti, *Geography*, **56**, 145–7.

Pollock, N. C. (1974) *Animals, Environment and Man in Africa*. University Press, Glasgow.

Popovic, V. (1973) *Tourism in East Africa*. Weltforum Verlag, Munich.

Prince Edward Island, Department of Tourism, Parks and Conservation (1975) *Tourism Employment Study: Prince Edward Island*. Charlottetown.

Quandt, R. E. (1970) *The Demand for Travel: Theory and Measurement*. Heath Lexington Books, Lexington, Mass.

Reid, N. J. (1967) Public view of wildlife, pp. 77–80 *Ecological Impact of Recreation and Tourism upon Temperate Environments*, IUCN Proceedings and Papers, New Series No. 7. Morges, Switzerland.

Rensberger, B. (1977) This is the end of the game, *New York Times Magazine*, **40**(3), 6 November, 38–43, 136–148.

Reynoso y Valle, A. and **De Regt, J. P** (1979) Growing pains: planned tourism development in Ixtapa-Zihuatanejo, pp. 111–34 in De Kadt, E. *Tourism – Passport to Development?* Oxford University Press, New York.

Ricciuti, E. R. (1976) Mountains besieged, *International Wildlife*, **6**(6), 24–35.

Richard, C. (1973) Tourism and public health, *South Pacific Bulletin*, **23**(1), 32–3.

Ritchie, J. R. and **Zins, M**. (1978) Culture as a determinant of the attractiveness of a tourist region, *Annals of Tourism Research*, **5**, 252–67.

Ritter, W. (1975) Recreation and tourism in the Islamic countries, *Ekistics*, **40**, 236, 56–9.

Rivers, P. (1973a) Tourist troubles, *New Society*, **23**, 539, 250.

Rivers, P. (1973b) No package deal, *New Society*, **25**, 566, 349–50.

Rivers, P. (1974) Unwrapping the African tourist package, *Africa Report*, **19**(2), 12–16.

Robinson, H. (1976) *A Geography of Tourism*. MacDonald & Evans, London.

Rodenburg, E. E. (1980) The effects of scale in economic development: tourism in Bali, *Annals of Tourism Research*, **7**, 177–96.

Rodriguez, F. (1978) 'Tourism and the environment': an interview by the World Travel Organization, *World Travel*, **143**, 49–51.

Roebuck, J. and **McNamara, P**. (1973) Ficheras and free lancers: prostitution in a Mexican border city, *Archives of Sexual Behaviour*, **2**, 231–44.

Ropponen, P. J. (1976) Tourism and the local population, pp. 104–9 in Economic Commission for Europe, *Planning and Development of the Tourist Industry in the E.C.E. Region*. United Nations, New York.

Rostron, J. (1972) The contribution of attitude studies to outdoor recreation planning, pp. 31–40 in Foster, H. D., and Sewell, W. R. D. (eds), *Resources, Recreation and Research*, Occasional Paper No. 13. Department of Geography, University of Victoria, British Columbia.

Rothman, R. A. (1978) Residents and transients: community reaction to seasonal visitors, *Journal of Travel Research*, **16**(3), 8–13.

Sadler, P. and **Archer, B. H**. (1974) *The Economic Impact of Tourism in Developing Countries*, Tourist Research Paper No. 4. Institute of Economic Research, University College of North Wales, Bangor.

Saglio, C. (1979) Tourism for discovery: a project in Lower Casamance, Senegal, pp. 321–35 in De Kadt, E. *Tourism – Passport to Development?* Oxford University Press, New York.

Samuelson, P. A. (1967) *Economics: An Introductory Analysis*. McGraw-Hill, London.

Sandelowsky, B. H. (1976) Functional and tourist art along the Okavango River, pp. 350–65 in Graburn, N. H. *Ethnic and Tourist Arts: Cultural Expressions from the Fourth World*. University of California Press, Berkeley.

Satchell, J. E. and **Marren, P. R**. (1976) *The Effects of Recreation on the Ecology of Natural Landscapes*. Council of Europe, Strasbourg.

Sayer, S. (1973) Wildland uplands of the South West, *Geographical Magazine*, **45**, 743–7.

Schaer, U. (1978) Traffic problems in holiday resorts, *Tourist Review*, **33**(2), 9–15.

Schadler, F. (1979) African arts and crafts in a world of changing values, pp. 146–56 in De Kadt, E. *Tourism – Passport to Development?* Oxford University Press, New York.

Schmoll, G. A. (1977) *Tourism Promotion*. Tourism International Press, London.

Schneider, H. (1976) Tourist development in Africa, *Afrika Spectrum*, **76**, 5–16.

Sewell, W.R.D. and **Burton**, I. (eds) (1971) *Perceptions and Attitudes in Resources Management*. Information Canada, Ottawa.

Sewell, W. R.D. and **Coppock, J. T**. (eds) (1977) *Public Participation in Planning*. Wiley, Toronto.

Sewell, W. R. D., Davis, J and **Ross, D. W**. (1961) *A Guide to Cost Benefit Analysis*. Queen's Printer, Ottawa.

Shaw, R. R. (1973) Tourism: cleaning up the air, *Proceedings of the Pacific Area Travel Association*. Kyoto, Japan, 227–30.

Shiviji, I. G. (ed.) (1973) *Tourism and Socialist Development*. Tanzania Publishing House, Dar-es-Salaam.

Shoard, M. (1974) Opening the countryside to the people, *Journal of Planning and Environmental Law*, May, 266–71.

Sigaux, G. (1966) *History of Tourism*. Leisure Art, London.

Simon, H. A. (1957) *Models of Man*. Wiley, New York.

Singer, M. (1968) The concept of culture, vol. 3, pp. 527–41 in Sills, D. (ed.),

International Encyclopaedia of the Social Sciences. Macmillan, New York.
Smith, V. (ed.) (1977) *Hosts and Guests: An Anthropology of Tourism*. University of Pennsylvania Press, Philadelphia.
Socher, E. (1976) No litter please on Everest, *Geographical Magazine*, **48**, 388.
Soo Ann, L. (1973) Tourism builds a smokeless economic base, *Proceedings of the Pacific Area Travel Association*. Kyoto, Japan, 207–9.
Spicer, E. H. (1968) Acculturation, vol. 3, pp. 21–5 in Sills, D. (ed.), *International Encyclopaedia of the Social Sciences*. Macmillan, New York.
Stansfield, C. A. (1971) The geography of resorts: problems and potentials, *Professional Geographer*, **13**, 164–6.
Stansfield, C. A. (1972) The development of modern seaside resorts, *Parks and Recreation*, **5**(10), 14–17, 43–6.
Stansfield, C. A. (1978) Atlantic City and the resort cycle: background to the legalisation of gambling, *Annals of Tourism Research*, **5**, 238–51.
Stansfield, C. A. and **Rickert, J. E**. (1970) The recreational business district, *Annals of Tourism Research*, **4**, 213–25.
Statistics Canada (1979) *Travel, Tourism and Outdoor Recreation: A Statistical Digest*. Minister of Industry, Trade and Commerce, Ottawa.
Sunday, A. A. (1978) Foreign travel and tourism prices and demand, *Annals of Tourism Research*, **5**, 268–73.
Sutton, W. A. (1967) Travel and understanding: notes on the social structure of touring, *International Journal of Comparative Sociology*, **8**, 218–23.
Swift, J. (1972) What future for African National Parks, *New Scientist*, **55**, 806, 192–4.
Symanski, R, (1974) Prostitution in Nevada, *Annals of the Association of American Geographers*, **64**, 357–77.
Talbot, N. (1974) A note on tourism in the West Indies, *Science and Society*, **38**, 347–9.
Tangi, M. (1977) Tourism and the environment, *Ambio*, **6**, 336–41.
Thomas, J, (1964) What makes people travel, *Asia Travel News*, August, 64–5.
Thomason, P., Crompton, J. L. and **Van Kamp, B**. (1979) A study of the attitudes of impacted groups within a host community toward prolonged staying tourist visitors, *Journal of Travel Research*, **18**(3), 2–7.
Thorsell, J. W. (1973) National parks in developing countries, *Agricultural and Forestry Bulletin*, **2**, 17–19.
Thuens, H. L. (1976) Notes on the economics of international tourism in developing countries, *Tourist Review*, **31**(3), 2–10.
Tourism and Recreation Research Unit, University of Edinburgh (1976–77) *Research Study into Provision for Recreation in the Highlands and Islands: Phase 1 – Areas Affected by Oil-related developments* (3 vols).
Travis, A. S. (1980) The need for policy action, pp. 79–97, in Organisation for Economic Co-operation and Development, *The Impact of Tourism on the Environment, Paris*.
Triantis, S. G. (1979) Economic impact of tourism and recreation in Muskoka, pp. 273–9 in Wall, G. (ed.) *Recreational Land Use in Southern Ontario*, Publication Series No. 13. Department of Geography, University of Waterloo, Ontario.
Turner, L. (1976) The international division of leisure: tourism and the third world, *World Development*, **4**, 253–60.
Turner, L. and **Ash, J**. (1975) *The Golden Hordes: International Tourism and the Pleasure Periphery*. Constable, London.
UNESCO (1976) The effects of tourism on socio-cultural values, *Annals of Tourism Research*, **4**, 74–105.
Urbanowicz, C. F. (1977a) Tourism in Tonga: troubled times, pp. 83–92 in Smith, V. (ed.) *Hosts and Guests: The Anthropology of Tourism*. University of Pennsylvania Press, Philadelphia.
Urbanowicz, C. F. (1977b) Integrating tourism with other industries in Tonga, pp. 89–93 in Farrell, B. H. (ed.), *Social and Economic Impact of Tourism on Pacific*

Communities. Centers for South Pacific Studies, University of California, Santa Cruz.
US News and World Report (1976) Jam-up at vacation spots, **80**(5), 26–8.
Van der Werff, P. (1980) Polarizing implications of the Pescaia tourist industry, *Annals of Tourism Research*,**7**, 197–223.
Vaughan, R. (1977a) *The Economic Impact of Tourism in Edinburgh and the Lothian Region*. Scottish Tourist Board, Edinburgh.
Vaughan, R. (1977b) *The Economic Impact of the Edinburgh Festival 1976*. Scottish Tourist Board, Edinburgh.
Vaughan, R. (1977c) in Leisure Studies Association Conference, *Tourism: A Tool for Regional Development*. Tourism and Recreation Research Unit, University of Edinburgh, 1977.
Vetter, F. (1975) Present changes in West German big city tourism, pp. 53–9 in Helleiner, F. (ed.), *Tourism as a Factor in National and Regional Development*, Occasional Paper No. 4. Department of Geography, Trent University, Peterborough, Ontario.
Wagar, J. A. (1964) *The Carrying Capacity of Wild Lands for Recreation*, Forest Science Monograph No. 7. Society of American Foresters, Washington, 3, 21.
Wagner, P. L. (1958) Remarks on the geography of language, *Geographical Review*, **48**, 86–97.
Wahab, S. E. (1975) *Tourism Management*. Tourism International Press, London.
Wahab, S. E., Crampon, J. and **Rothfield, L**. (1976) *Tourism Marketing*. Tourism International Press, London.
Wall, G.(1971: Car-owners and holiday activities, pp. 97–111 in Lavery, P. (ed.), *Recreational Geography*. David and Charles. Newton Abbot.
Wall, G. (1975) Form and function in British seaside resorts, *Society and Leisure*, **7**, 217–26.
Wall, G. (1978) Competition and complementarity: a study in park visitation, *International Journal of Environmental Studies*, **13**, 35–41.
Wall, G. and **Knapper, C.** (1981) *Tutankhamun in Toronto*, Publication Series No. 17. Department of Geography, University of Waterloo, Ontario.
Wall, G. and **Maccum Ali, I**. (1977) The impact of tourism in Trinidad and Tobago, *Annals of Tourism Research*, **5**, 43–50.
Wall, G. and **Sinnott, J**. (1980) Urban recreational and cultural facilities as tourist attractions, *Canadian Geographer*, **24**, 50–9.
Wall, G. and **Wright, C**. (1977) *The Environmental Impact of Outdoor Recreation*, Publication Series No. 11. Department of Geography, University of Waterloo, Ontario.
Waters, S. R. (1966) The American tourist, *Annals of the American Academy of Political and Social Sciences*, **368**, 109–18.
Waters, S. R. (1974) The impact of the energy crisis on tourism in the third world countries, *Travel Research Journal*, 23–7.
Watson, A. (1967) Public pressures on soils, plants and animals near ski-lifts in the Cairngorms, pp. 38–45 in Duffey, E. (ed.), *The Biotic Effects of Public Pressures on the Environment*, Monks Wood Experimental Station Symposium No. 3. Nature Conservancy, Monks Wood, Abbots Ripton, Huntingdon.
Waugh, R. E. (1962) *The American Traveller: More Darkness than Light?* Bureau of Business Research, University of Texas, Austin.
Westhoff, V. (1967) The ecological impact of pedestrian, equestrian and vehicular traffic on vegetation, pp. 219–23 in International Union for the Conservation of Nature and Natural Resources, *Ecological Impact of Recreation and Tourism on Temperate Environments*, IUCN Proceedings and Papers, New Series No. 7. Morges, Switzerland.
White, J. (1967) *History of Tourism*. Leisure Art, London.
White, P. E. (1974) *The social impact of tourism on host communities: a study of language change in Switzerland*, Research Paper No. 9. School of Geography, Oxford University.

Willard, B. E. and **Marr, J. W.** (1970) Effects of human activities on alpine tundra ecosystems in Rocky Mountain National Park, Colorado, *Biological Conservation*, **2**, 257–65.

Willard, B. E. and **Marr, J. W.** (1971) Recovery of alpine tundra under protection after damage by human activities in the Rocky Mountains of Colorado, *Biological Conservation*, **3**, 181–90.

Willis, F. R. (1977) Tourism as an instrument of regional economic growth, *Growth and Change*, **8**(2), 43–7.

Wilson, D. (1979) The early effects of tourism in the Seychelles, pp. 205–36 in De Kadt, E. *Tourism – Passport to Development?* Oxford University Press, New York.

Wimberly, G. J. (1977) Resort hotels: the challenge of gracious design, *Cornell Hotel and Restaurant Administration Quarterly*, **18**, 24–32.

Wolf, C. P. (1977) Social impact assessment: the state of the art updated, *SIA Newsletter*, **29**, 3–23.

Wolfson, M. (1967) Government's role in tourism development, *Development Digest*, **5**(2), 50–6.

Wolski, J. (1977) Polish resort-centres of health tourism, *Proceedings of the Association Internationale d'experts Scientifiques du Tourisme* (Berne, Switzerland), **18**, 203–5.

World Bank (1972) *Tourism–Sector Working Paper*. Washington, DC.

World Council of Churches (1970) *Leisure-Tourism: Threat and Promise*. Geneva, Switzerland.

World Tourism Organization (1978) *Economic Review of World Tourism*. Geneva, Switzerland.

Young, G. (1973) *Tourism: blessing or blight?* Penguin, Harmondsworth.

Young, G. (1975) Tourism: blessing or blight?, *Development Digest*, **13**, 43–54.

Index

Subjects

Authors

Locations